OCS Study
MMS 2005-029

Coastal Marine Institute

Modeling Structure Removal Processes in the Gulf of Mexico

I0476161

Authors

Mark J. Kaiser
Dmitry V. Mesyanzhinov
Allan G. Pulsipher

May 2005

Prepared under MMS Contract
1435-01-99-CA-30951-85246
by
Center for Energy Studies
Louisiana State University
Baton Rouge, Louisiana 70803

U.S. Department of the Interior
Minerals Management Service
Gulf of Mexico OCS Region

Cooperative Agreement
Coastal Marine Institute
Louisiana State University

DISCLAIMER

This report was prepared under contract between the Minerals Management Service (MMS) and Louisiana State University's Center for Energy Studies. This report has not been technically reviewed by MMS. Approval does not signify that the contents necessarily reflect the view and policies of the Service, nor does mention of trade names or commercial products constitute endorsement or recommendation for use. It is, however, exempt from review and compliance with MMS editorial standards.

REPORT AVAILABILITY

Extra copies of the report may be obtained from the Public Information Office (Mail Stop 5034) at the following address:

U.S. Department of the Interior
Minerals Management Service
Gulf of Mexico OCS Region
Public Information Office (MS 5034)
1201 Elmwood Park Boulevard
New Orleans, Louisiana 70123-2394
Telephone Number: 1-800-200-GULF
1-504-736-2519

CITATION

Suggested citation:

Kaiser, M.J., D.V. Mesyanzhinov, and A.G. Pulsipher. 2005. Modeling structure removal processes in the Gulf of Mexico. U.S. Dept. of the Interior, Minerals Management Service, Gulf of Mexico OCS Region, New Orleans, La. OCS Study MMS 2005-029. 133 pp.

ACKNOWLEDGMENTS

The advice and critical comments of Kristen Strellec, Tommy Broussard, Harry Luton, Jeff Childs, and Sarah Tsoflias are gratefully acknowledged. Editorial assistance provided by Ric Pincomb and Versa Stickle is also acknowledged. This paper was prepared on behalf of the U.S. Department of the Interior, Minerals Management Service, Gulf of Mexico OCS Region and has not been technically reviewed by the MMS. The opinions, findings, conclusions, or recommendations expressed in this paper are those of the authors, and do not necessarily reflect the views of the Minerals Management Service. Funding for this research was provided through the U.S. Department of the Interior and the Coastal Marine Institute, Louisiana State University.

ABSTRACT

At the end of 2003, there were nearly 4,000 structures in the federal waters of the Gulf of Mexico (GOM) associated with hydrocarbon production: 2,175 active (producing) structures, 1,227 idle (non-producing) structures, and 505 auxiliary (never-producing) structures. Since 1947, when production in the GOM first began, over 2,200 structures have been removed from federal waters, and over the past decade, 125 structures on average have been removed annually. The purpose of this report is to describe the operational aspects of removal processes in the GOM and to develop a production-based model to forecast the removal of offshore structures.

In Chapter 1, a statistical description of the explosive removal process is presented. The influence of factors such as water depth, planning area, configuration type, and structure age upon the application of explosive removal methods is described. Estimates for the number of structures that are expected to be removed from the GOM over a 25-year time horizon are forecast using a heuristic life expectancy and probabilistic removal model.

In Chapter 2, the factors involved in the decision to use a specific severance technique are described, and the probability that a structure will be removed with explosive technology is quantified. A simple predictive model of the decision to use explosive methods is also developed. An empirical analysis of GOM structures removed between 1986-2001 provide the historic data required to compute the probability of an explosive removal and to estimate binary choice models for severance selection. Binomial logit and probit models of severance selection are constructed to establish the relationship between structure attributes and the probability that a particular severance technique will be employed.

In Chapter 3, four models of abandonment timing decisions are developed, ranging from a resource-based forecast to a risked, net present value approach. Meta-modeling simulation is employed to construct functionals that describe how the age of the structure upon abandonment is related to the system parameters. The sensitivity of the results to differences in the model assumptions, and the practical matter of whether sophisticated models are, in fact, more accurate than simple forecasting techniques, is discussed. A generic field development scenario is used to illustrate the decommissioning timing models.

In Chapter 4, the economic limit of offshore structures is estimated using historic data from GOM structures removed over the past two decades. This is the first time that threshold limits of production near abandonment have been investigated and quantified.

In Chapter 5, a production-based model to forecast removal rates and costs of offshore structures is presented. A stochastic decline model is used to forecast production, and in conjunction with estimates of the economic limit, is used to determine the time that a structure is abandoned. The expected time a structure is removed is based on federal regulatory requirements which determine the latest possible removal scenario. A description of the modeling framework and results are presented, along with a discussion of the limitations of analysis.

TABLE OF CONTENTS

TABLE OF CONTENTS (continued)

LIST OF FIGURES

LIST OF TABLES

LIST OF TABLES (continued)

CHAPTER 1: EXPLOSIVE REMOVALS OF OFFSHORE STRUCTURES

1.1. Introduction

The Outer Continental Shelf (OCS) of the U.S. Gulf of Mexico (GOM) is one of the most highly developed and mature basins in the world. Over the last 50 years, the oil and gas industry has installed over 6,000 structures and 33,000 miles of interconnecting pipelines in the gulf waters. Today, there are about 4,000 active structures installed in federal[1] water ranging from less than 10 feet to over 7,000 feet. There are also a few thousand structures in state waters off the coast of Louisiana and Texas, almost all of which are small and installed in less than 35 feet of water.

Structures need to be constructed, delivered, installed, and equipped prior to production, operated and serviced during production, and then eventually decommissioned and removed after production. Each of these activities has both a direct and indirect impact on the communities in which the service facilities and manufacturing operations are located, and hence induce a "spill-over" effect on the economic growth of regions which serve the development. An entire industry has been built in the GOM around installing production equipment and structures, servicing those structures (maintenance, repairs, supply), and then removing the structures when production ceases.

During the life of a lease, the leaseholders apply for permits to place structures on the seafloor to aid in drilling, development, and production operations. Near the end of the economic life of the lease, when the structures have been fully depreciated and reserves depleted, the structures represents a financial and operational liability, and at this point in time a decision is made to abandon. Within one year of lease termination, the Minerals Management Service (MMS) requires that the lessees remove all structures to a depth of 15 feet below the mudline and that the site be returned to prelease conditions. Although multiple techniques may be used to sever the structural components, they are generally categorized as either explosive or nonexplosive methods.

Operators wishing to remove an OCS platform or facility are required to submit a structure removal permit application to MMS for technical review and the preparation of an environmental assessment (EA) under National Environmental Policy Act (NEPA) guidelines. Prior to mobilization, additional permits are required for well abandonment (temporary or permanent) and/or pipeline decommissioning to ensure that all of the infrastructure components to and from the structure are secured. Removal operations proposing explosive severance are currently subject to the terms and conditions of a programmatic Biological Opinion (BO)/ Incidental Take Statement (ITS) issued by the National Oceanographic and Atmospheric Administration's Fisheries Service (NOAA Fisheries) under an Endangered Species Act (ESA) Consultation with MMS. If an operator proposes any activities that fall outside of the BO/ITS severance criteria (e.g., 50-lb maximum charge weight, cut depth, 900 msec detonation staggering, etc.), a site-specific ESA Consultation and new BO/ITS will be required.

---

[1] Federal jurisdiction in the OCS varies with the Gulf state: Florida and Texas have an extended nine nautical mile state jurisdiction, while Alabama, Louisiana, and Mississippi have the standard three nautical mile state jurisdiction.

The NOAA Fisheries Service currently assigns observers to every OCS structure removal operation using explosive charges >5 lb. A pre-blast aerial survey is conducted immediately prior to the explosive detonation using a helicopter with a NMFS observer on board. If marine mammals or sea turtles are found within 1,000 yards of the structure, the detonation is delayed until the area is clear. A post-blast aerial survey is conducted after the explosives are detonated to assess the impact to the marine life. Underwater detonations have the potential to harass, injure, or kill marine mammals and sea turtles; however, since introduction of the NOAA Fisheries' Platform Removal Observer Program (PROP) in 1986, only two sea turtles have been killed and three turtles have been injured as a result of explosive severance.

The purpose of this chapter is to provide a statistical description of structures that have been removed in the GOM and the manner of their removal. The influence of factors such as water depth, planning area, configuration type, and structure age will be examined, and the relationship of these factors with explosive removals will be discussed. Estimates for the number of structures that are expected to be removed from the GOM over a 25-year time horizon are forecast by configuration type, water depth, and planning area categorization over 5-year time blocks beginning from the year 2002. The result of the models and a description of the limitations of the analysis is then presented. Conclusions complete the chapter.

1.2. Statistical Description of Structure Removals

1.2.1. Notation: The GOM is partitioned according to protraction area, water depth, and planning area categories. The three planning areas which divide the GOM are denoted the Western, Central, and Eastern Gulf of Mexico (WGOM, CGOM, EGOM). See Figure A.1. Each planning area is subdivided into smaller regions, called protraction areas, which in turn are divided further into numbered blocks. Each block is designated by a number and is normally a nine square mile area consisting of 5,760 acres. A single block is the smallest unit that can be leased for oil and gas exploration on the OCS. The water depth categorization applied in MMS resource assessment evaluations,

$$W = \{W_1, \ldots, W_5\} = \{0\text{-}200\text{m}, 201\text{-}800\text{m}, 801\text{-}1600\text{m}, 1601\text{-}2400\text{m}, 2400^+\text{m}\},$$

is too broadly defined for platform removal studies since nearly all (96%) of the structures removed in the GOM to date have been within a 0-60 meters water depth range. To examine the use of explosive methods as a function of water depth a finer level of disaggregation needs to be employed. A partition is selected that decomposes the 0-200 meters category into a 0-60 meters and 61-200 meters category, and then a further partition of the 0-60 meters category into subcategories is adopted. Eleven subcategories within the 0-60 meters water depth range classified according to feet are applied, and we shall transition between the distance measures as convenience dictates. The first five subcategories employ a ten feet range, and then from 50-200 feet, a 25 feet range is employed:

$$W = \{W_1, \ldots, W_4\} = \{0\text{-}10, 11\text{-}20, 21\text{-}30, 31\text{-}40, 41\text{-}50, 51\text{-}75, 76\text{-}100, 101\text{-}125, 126\text{-}150, 151\text{-}175, 176\text{-}200, 201\text{-}656, 657\text{-}2624, 2624^+\text{ft}\}.$$

The GOM planning areas are denoted by

$$P = \{ P_1, P_2, P_3 \} = \{\text{WGOM, CGOM, EGOM}\},$$

and since the Eastern GOM has seen only a very small level of activity, this planning area will not be considered further. Since the water depth and planning area schemes are disjoint, the two categories can be combined using a Cartesian product as follows:

$$W \times P = \{\Gamma_{i,j} = (W_i, P_j) \mid i = 1,\ldots,14; \ j = 1,2\},$$

where $\Gamma_{i,j}$ denotes the water depth and planning area category indexed by i and j; e.g., $\Gamma_{4,2}$ denotes the 31-40 feet water depth range in the Central GOM.

Structures can be classified through their attributes such as configuration type and age upon removal. Configuration type is described using four categories as follows:

$$\{ T_1, T_2, T_3, T_4 \} = \{\text{caissons, well protectors, fixed, floating}\}.$$

The minimum structure for offshore development of a well is a caisson, a cylindrical or tapered tube enclosing the well conductor. A small deck is sometimes provided above the wellhead, but no facilities are provided except possibly navigational aides and a small crane (Figure A.2). Structures that provide support to one or more wells drilled with a mobile drilling rig are normally referred to as well protectors. Well protectors are sized to fit within the drilling slot of a mobile drilling rig, and are usually 3- or 4-piled structures with minimum decks and production facilities (Figure A.3). Production from caissons and well protectors is usually sent to a production platform for treating. Well protectors and other fixed platforms are designed with a jacket, a three-dimensional welded frame of tubular members, used as a guide for driving piles through its legs. Fixed platforms include drilling, production, drilling/production, and auxiliary platforms (Figure A.4). Depending on the design and construction requirements and constraints, the number of piles of a fixed platform can vary from three to eight or more and can be as small as 24 inches or as large as 96 inches. Four-pile and 8-pile fixed platforms are the most common structures in the GOM.

The age of the structure upon removal is grouped according to

$$\{ A_1, A_2, A_3, A_4 \} = \{\text{0-10, 11-20, 21-30, 30}^+ \text{ years}\}.$$

The number of structures removed from the water depth and planning area region $\Gamma_{i,j}$ over the time interval $(t-1,t)$ is specified in terms of configuration type and age as follows:

$R(\Gamma_{i,j}, T_k, t)$ = Number of structures removed from region $\Gamma_{i,j}$ of type T_k in year t,

$R(\Gamma_{i,j}, A_l, t)$ = Number of structures removed from region $\Gamma_{i,j}$ that fall within age group type A_l in year t,

3

$R(\Gamma_{i,j},\ T_k,\ A_l,\ t) =$ Number of structures removed from region $\Gamma_{i,j}$ of type T_k that fall within age group A_l in year t.

The number of structures removed using explosive methods is denoted by the subscript E; e.g.,

$R_E(\Gamma_{i,j},\ T_k,\ t) =$ Number of structures removed from region $\Gamma_{i,j}$ of configuration type T_k using explosive techniques in year t.

The percentage of structures of a given classification that are removed through explosive technology is computed as the ratio of $R_E(\cdot)$ to $R(\cdot)$; e.g., the percentage of structures of configuration type T_k removed through explosive technology in year t is computed as

$$p_E(\Gamma_{i,j},T_k,t) = \frac{R_E(\Gamma_{i,j},T_k,t)}{R(\Gamma_{i,j},T_k,t)},$$

and in most cases time will be "integrated out" of the data set:

$$p(\Gamma_{i,j},T_k) = \frac{\sum_t R_E(\Gamma_{i,j},T_k,t)}{\sum_t R(\Gamma_{i,j},T_k,t)}.$$

Percentage applications must be employed cautiously, however, since if the number of elements in the set $R(\cdot)$ or $R_E(\cdot)$ is "small," then $p_E(\cdot)$ cannot be considered a reliable statistic. For instance, if there are less than a dozen elements in a given set, then one cannot assign much confidence to the values as being "representative" of conditions in the region. The tables of summary statistics present raw data as well as the percentage values to convey this information.

The total number of structures removed from the water depth and planning area region $\Gamma_{i,j}$ is denoted by $R(\Gamma_{i,j})$ and is equal to any complete summation of the decomposed data over the universe of the partition:

$$R(\Gamma_{i,j}) = \sum_t \sum_{k=1}^{4} R(\Gamma_{i,j},T_k,t) = \sum_t \sum_{l=1}^{4} R(\Gamma_{i,j},A_l,t) = \sum_t \sum_{k=1}^{4} \sum_{l=1}^{4} R(\Gamma_{i,j},T_k,A_l,t).$$

1.2.2. Structure Installation and Removals by Water Depth: Information on offshore structures in federal waters was obtained from the U.S. Minerals Management Service. The MMS updates its database on a periodic basis as new information is made available, and so there is always a time lag between when the data is reported and entered into the database and when it is analyzed. The structure data employed in this chapter was current through November 2001. The total number of structures installed and removed since 1947 as a function of water depth and planning area is depicted in Table A.1. Structures are defined to include all caissons, well protectors, fixed platforms, and floating configurations.

4

Nearly 6,000 structures have been installed in the GOM through the year 2001 and one-third of these structures have now been removed. The vast majority of installations and removals have been in shallow water: 90% of all structures installed in the GOM and 96% of all the removals have been in less than 200 feet (60 meters) of water. Within the 0-200 feet category, 36% of all the structures that have been installed through the year 2001 have been removed, while only 14% of structures beyond 200 feet have been removed. Activity levels vary widely as a function of water depth.

The average annual number of structures installed and removed per water depth and planning area category over a 5-year (1996-2001) and 10-year (1991-2001) time horizon is depicted in Table A.2 and Table A.3, respectively. The value of the average annual number of installations and removals is surprisingly robust over the 5- and 10-year horizon in the sense that the mean and standard deviation of the installation and removal rates do not change appreciably. On the other hand, activity levels are highly uncertain throughout most of the water depth categories, and so the normal statistical interpretation bounding the mean through a confidence interval employing one- or two-standard deviations should also be approached cautiously.

The CGOM and WGOM planning areas exhibit significantly different activity levels. The number of structures installed in the CGOM is roughly five times WGOM activity, and a similar level of activity governs the removal rates. In shallow waters, structure removal rates are comparable to installation rates across planning area. In deep waters, structure installations dominate removals. The historic magnitude of installation and removal activity is also clearly dominated by shallow water activity. To date 82% of all WGOM structures and 91% of all CGOM structures have been installed in less than 200 feet of water. In terms of structure removals, 91% of all WGOM removals and 96% of all CGOM removals have been in 200 feet of water or less.

1.2.3. Age Distribution of Active and Removed Structures: In Table A.4 the percentage of active structures that fall within each age category is depicted. Infrastructure in the GOM is aging and this is clearly indicated among all configuration categories, especially within the CGOM region where nearly 40% of the well protectors and over a third of the fixed platforms are over 30 years old.

In Table A.5 the average age of structures removed from the GOM is depicted. Structures in the WGOM tend to be removed, on average, earlier than their counterparts in the CGOM, which could be due to smaller field size, faster production rates, or other geologic-based conditions; e.g., most fields in the WGOM are gas fields which exhibit a quick depletion rate. Caissons in the CGOM are removed after about 16 years of service while caissons in the WGOM have a significantly shorter lifespan of seven years. Observe that the standard deviation values in all cases are greater than 50% of the value of the mean, and so it is clear that there is wide variability in structure removal ages within categories and across water depth and configuration type. Also, with the exception of well protectors in the WGOM, there is not a significant difference in the average age of removal across water depth categories.

1.2.4. Structure Removals by Configuration Type and Method of Removal: The application of explosive techniques varies widely throughout Gulf waters. In Table A.6, the number of structures removed $R = R(\Gamma_{i,j})$ and the number removed by explosive techniques ($R_E = R_E(\Gamma_{i,j})$) are shown as a function of water depth and planning area beginning from 1986. Although multiple techniques may be used to sever conductors and piling, severing is usually categorized as either explosive or nonexplosive. If explosives are used in any amount and at any stage of the decommissioning project, then the method is considered explosive. Beginning in 1986 companies planning to remove offshore structures with explosives were required to obtain a permit from the MMS, and hence only data from this period of time onward is available. The data set represents about 80% of the total structure removals to date.

The percentage of structures removed using explosive techniques is calculated as

$$p_E = \frac{R_E(\Gamma_{i,j})}{R(\Gamma_{i,j})}.$$

The percentage values depicted need to be interpreted carefully, however, since the values depend upon the selection of the water depth categories employed. An additional problem in interpreting the value of p_E is that the percentage calculation may be based on only a handful of data, and in such circumstances, one cannot assign much confidence to the values as being "representative" of conditions in the region. This is particularly a problem throughout the shallow water (0-40 feet) and deepwater (657-2,624 feet) categories of the WGOM where only a few structures have been removed. With these exceptions noted, however, there does not appear to be a significant difference between the application of explosive techniques over the WGOM and CGOM planning area, which is quite reasonable considering there is no rational reason why explosive techniques would be different across planning area unless the structure types, age[2], or year of removal are dramatically different. The data in Table A.6 supports the assertion that planning area dependence on p_E is weak, and so we can aggregate over planning area and consider the application of explosive removals throughout the GOM as representative of either the WGOM or CGOM planning area.

The description of explosive removals across the GOM as a function of configuration type is depicted in Table A.7. It is apparent from Table A.7 that the choice of removal method depends to some extent on the configuration type of the structure, but there are *no* observable trends *within* the 0-200 feet category for any of the configuration types. It is also difficult to explain the variability that does exist, and most probably, the variation of p_E with water depth is due to "noise" that cannot be detected. Recall that structure removal decisions are usually based on a few factors that are mostly unobservable: cost, safety, risk of failure, and technical feasibility, combined with a wide variety of structure-, site-, and company-specific criteria. The variation that exists in $p_E(W_i, T_k)$ across water depth and configuration types leads us to conclude that the best indicators of p_E are aggregate measures across broad water depth categories that do not differentiate between planning area.

---

[2] In fact, as mentioned previously (recall Table A.5), there is a difference in the average age of structures upon removal across planning area, and to the extent that a younger structure is *more* likely to be removed with nonexplosive technology, we would suspect WGOM structures to have a slightly *lower* probability of being removed using explosives as shown in the aggregated categories $p_E(P_1)$ and $p_E(P_2)$ at the bottom of Table A.6.

Using the categorization shown at the bottom of Table A.7, observe that caissons are the most commonly removed structure using nonexplosive methods, and well protectors and fixed platforms, if removed using nonexplosive techniques, are more commonly performed in shallow waters. Caissons have an equal chance of being removed with either explosive or nonexplosive methods, and well protectors and fixed structures realize a greater chance of an explosive removal. As the water depth increases the chance of using explosives also increase across all configuration types. The percentage values depicted for explosive removals for well protectors in the 61-200 meters water depth range is slightly suspect, however, since it is based on only six data points. Thus far, no caissons, well protectors, or fixed structures have been removed in water depth greater than 200 meters, and the two semisubmersibles that have been removed in this water depth range are included for completeness.

1.2.5. Structure Removals by Year and Configuration Type: The number of structures removed by configuration type by year is shown in Table A.8 across all water depths in the Gulf of Mexico. There are no noticeable trends in the removal rates across time except caissons and fixed structures typically compete for the greatest number of removals in any given year. The percentage values p_E can be considered a stochastic process, but it is preferable to "average out" the time variability by aggregating the $R_E(\cdot)$ and $R(\cdot)$ values and calculating

$$p_E(T_k) = \frac{\sum_t R_E(T_k, t)}{\sum_t R(T_k, t)},$$

as shown in the last row of Table A.8. The variability of p_E across time for a given configuration class can be explained to some extent through the age of the structure and the water depth.

1.2.6. Structure Removals by Age, Water Depth, and Configuration Type: Structures that have been removed from the GOM according to planning area and age upon removal are depicted in Table A.9. All structure types are aggregated within the same category and it is clear that a significant variation exists across planning areas. More than 90% of all WGOM structures are removed within 20 years of their installation – indicating small reserves, quick production, poor geologic prospects, or a combination of all these factors. For the most part, structures in the GOM are removed several years after they reach the end of their economic life. A few structures are removed early because of structural damage (collision with barge, hurricane event, etc.), with fewer still removed because of fatigue. Structures are removed near the time when they are no longer economic, and this is not (normally) constrained by the design life of the structure. At the opposite end of the spectrum is the relatively long life of many CGOM fields: 15% of all CGOM structures for instance were at least 30 years old upon removal.

In Table A.10 the number of structures removed using explosives is depicted along with the percentage of explosive removals categorized according to age. Examine the percentage of explosive removals shown on the right side of Table A.10. As the age of a structure increases, so does the frequency that explosive methods will be employed. It is interesting to note that when the data is aggregated according to age upon removal, WGOM structures have a greater likelihood of an explosive removal relative to CGOM structures.

To examine the features of water depth and structure age upon removal method, structure data was aggregated and then classified as shown in Table A.11 and Table A.12. Table A.11 depicts the number of structures removed as a function of water depth and age upon removal, and it is clear that the majority of structures removed from both water depth categories are within 20 years of their installation date. The data in Table A.12 are more interesting, however, since the general trends observed earlier hold here with the same caveats: the percentage of structures removed using explosive methods increase as a function of age upon removal for the 0-60 meters category and is dominated by the application of explosive removals in the 61-200 meters water depth category. The number of structures in the 61-200 meters group, however, especially for the 21-30 and 30+ age categories, is too small to draw meaningful conclusions.

The general trends observed in Table A.7 for the application of explosive techniques also apply to individual configuration type and water depth categories as shown in Table A.13 and Table A.14. In Table A.13, observe that across all configuration types, the use of nonexplosive methods is most common in the 0-10 year category, and as the age of the structure increases, so does the likelihood that explosive methods will be applied. In Table A.14, the percentage of structures removed using explosives as a function of water depth, age upon removal, and configuration type is presented. Blank entries indicate that no structures within the given categorization were removed.

1.3. A Life Expectancy Model of Platform Removal Processes

1.3.1. A Structure Has at Least Five Lives: An offshore structure is an economic investment that has at least five distinct "lives": (1) the physical life, (2) the service life, (3) the depreciation life, (4) the design life, and (5) the economic life.

The physical life of a structure is the period of time over which the investment is actually used, while the service life is the period of time over which the structure is held for a particular purpose or level of service. The physical life of a structure is not necessarily identical with the service life, since a structure on a lease may cease to produce hydrocarbons but is maintained in service to be removed and decommissioned at a later date.

The depreciation life of a structure is the period of time over which the investment is depreciated on the operators accounting books. The depreciation period starts each time the structure is placed in service by a new owner and ends when (a) the structure is decommissioned, (b) all of the allowable depreciation is deducted, or (c) the structure is no longer used for business purposes.

The design life of a structure depends upon the expected field size and operator development plan. In the early years of offshore development, structures were typically over-designed to compensate for the uncertainties of a hostile and unknown environment. Today, operators have greater experience balancing the cost and risk of field development, and the design life of structures usually far exceed the actual field life. The design standards for platforms are specified according to design loads for specific oceanographic criteria, including wave directionality, current velocity, wave period, and wind speed. Structures in the GOM are designed to withstand a 100-year return period for hurricane wind, wave, and current environment.

The economic life of a structure is defined as the time at which the production cost of the structure is equal to the production revenue. At the time a structure reaches its economic limit, production will cease and operations will be abandoned. A lease may reach its economic limit prematurely when hydrocarbon prices are in a depressed price-demand state, but if the operator believes stronger prices will prevail in the future, then an abandonment decision is likely to be postponed until the operator can no longer sustain operating losses.

1.3.2. Sources of Uncertainty: Decommissioning represents a liability as opposed to an investment, and the pressure for an operator to decommission a structure is not nearly as strong as installation activities. There are usually no commercial incentives for early removal and operators have no incentive to "fast track" decommissioning unless pushed by regulatory time limitations.

Several sources of uncertainty impact decommissioning decision making:

- Geologic uncertainty,

- Production uncertainty,

- Price uncertainty,

- Investment uncertainty,

- Technological uncertainty, and

- Strategic uncertainty.

Production engineers estimate the reserve potential of a field based on geologic and geophysical data and then use this information to design the capacity of the structure and optimize the production schedule. Production profiles are used as a guideline to expected removal times since investment activity can dramatically alter the form of the production curve as well as the recoverable reserves. Hydrocarbon price, technological improvements, and demand-supply relations impact the revenue of the lease which also impact investment planning. When the time arrives that the cost to operate a lease (maintenance, operating personnel, transportation, fuel, insurance etc.) outstrips the income from production, the structures on the lease exist as liabilities instead of assets, and a decision is made to divest the property or abandon the structure subject to the strategic objectives of the operator. Strategic objectives are generally unobservable, nonquantifiable, and vary over time, region, and operator, further exacerbating the capability of forecast models.

1.3.3. Removal and Severance Models:

Life Expectancy Removal Model

The removal date of a structure is estimated through the relation

$$r(s) = i(s) + a(\Gamma) + k\sigma(\Gamma),$$

where,

$r(s)$ = Year of removal of structure s,
$i(s)$ = Year of initial production of structure s,
Γ = Classification category,
$a(\Gamma)$ = Average age upon removal for structure $s \in \Gamma$,
$\sigma(\Gamma)$ = Standard deviation of the age statistic.

The value for $a(\Gamma)$ and $\sigma(\Gamma)$ is defined according to configuration type, water depth and planning area, as shown in Table A.3. The value of k is user-defined.

The primary assumption of the model is that the historical characteristics of structures can be used to reasonably predict the removal trends of "similar" active structures, where "similarity" is defined for structures that fall within the same general classification category. The assumption is restrictive but is considered an acceptable first-order approximation.

The removal model adopts the approach taken by the National Research Council (NRC) 1985 report, where values for $a(\Gamma)$ were estimated as follows: "Smaller structures in shallow waters, such as caissons and well protectors, tend to be removed after 20-25 years; larger structures with more wells, such as 4- and 8-pile platforms, have a useful life of 25-30 years, and larger structures in deepwater should have a useful life of at least 30 years." The NRC heuristic approach is re-calibrated by computing the values of $a(\Gamma)$ and $\sigma(\Gamma)$ based on historic data, and then selecting k as a user-defined variable.

In Model I, set $k = 1$ and compute $r(s)$. If $r(s) \geq 2002$, then "accept" the removal time of structure s; otherwise, set $k = 3$. In Model II, the smallest integer value of k is determined such that $r(s) \geq 2002$, and for this value "accept" the removal time of the structure. Model I and Model II ensure that all installed structures will be removed based on their installation date and average age of removal plus a perturbation term. Model I presents a slow removal scenario; Model II presents an accelerated removal schedule.

Explosive Severance Model

The decision to employ explosive techniques in cutting operations depends upon a number of factors, and to the extent that these variables can be proxied by configuration type, water depth, and age upon removal, the probability that a structure will be removed using explosive techniques is written as $p_E(s)$. Structure s belongs to category Γ and is estimated to be removed at the time $r(s)$. Since the age of the structure being removed is known when $r(s)$ is "accepted," the value of $p_E(s)$ is extracted from Table A.14 to determine the probability the structure will be removed with explosives.

1.3.4. Model Results: The forecast output predicts the number of structures expected to be removed using explosive technology categorized by configuration type, water depth, and planning area across 5-year time blocks, where the block $200X- 200(X+4)$ is interpreted as January 1, $200X$ − December 31, $2000(X+4)$. A summary of the number of active structures expected to be removed with explosives is depicted in Table A.15 and Table A.16. A reasonable planning level suggests that between 94 and 159 structures per year will be removed with explosives in the short-term future. Structure composition indicates that major structures will play an increasingly important role both in terms of the absolute number of structures that will need to be removed as well as the expected cost of removal.

1.3.5. Model Assumptions: All removal forecasts need to be viewed relative to their structural framework. The assumptions that provide the framework to perform a forecast also, to varying extent, limit the interpretation of model results. Since operator behavior is too complex to model on an aggregate basis without the use of production profiles or private information (e.g., nomination schedules, leasehold operational cost, field development plans, strategic objectives, etc.), all non-production based forecasts are considered to have comparable levels of uncertainty. Within the class of non-production based models, the magnitude of the uncertainty cannot be mitigated through the selection of more advanced methodologies. In fact, more "advanced" approaches merely disguise and shift the uncertainty rather than actually reduce or mitigate it. Heuristic methods have some advantage over sophisticated procedures in such an environment relative to ease of implementation and focus on the model drivers. On the other hand, heuristic procedures are also rather arbitrary, and it is often desirable to investigate more advanced techniques to refine and improve the model structure.

A life expectancy and probabilistic removal model is considered an appropriate *first-order* approximation to predict removal times. Better models exist, but these models are considerably more difficult to construct and are subject to their own sources of uncertainty.

The primary assumptions employed in the life expectancy and explosive severance model are as follows:

A1. Structures are differentiated according to configuration type, water depth, and planning area, and are removed based on the time history of their installation and their estimated age at the time of decommissioning;

A2. The removal of structures installed in the future are not considered;

A3. The probability that an active structure is removed with explosives is assumed to depend solely on the configuration type, water depth, and age upon removal;

A4. The operating cost and production revenue associated with a structure, complex, leasehold, or field is not considered;

11

A5. Operator-specific conditions such as investment strategy, field development options, removal scheduling, regulatory constraints, and divestiture opportunity that may influence structure removal times are not considered.

1.3.6. Limitations of the Analysis: Since all active structures are differentiated on a configuration type, water depth, and planning area basis, and are removed according to the time history of their installation (A1), the categorization and removal relation ensures a structured methodology.

Most structures that will be removed over the next two decades will come from the population of existing structures. Structures expected to be installed in the future may need to be removed at a time which overlaps with the current time horizon, but most structure removals associated with future installations is expected to occur outside or near the end of the time horizon of the current forecast. Assumption (A2) is thus considered a reasonable assumption over a medium term horizon.

The life expectancy removal relation is regarded as a simple first-order heuristic to approximate the expected removal time. The removal relation is a simple, non-production based model, and in view of the sources and magnitude of uncertainty governing the problem, is considered appropriate. In the absence of company-level economic criteria or engineering estimates of field life, the values of the parameters used with this relation are based on historic data for elements classified within reasonably homogeneous categories.

The probability of an explosive removal is characterized using aggregate statistics categorized according to configuration type, water depth, and age upon removal. More advanced regression-based methodologies could be examined to predict the removal method as a function of these variables, but data limitations are expected to constrain the viability of this approach. Assumption (A3) is considered appropriate relative to the constraints of the model and the manner in which decisions are made by operators.

It is desirable to build a model that mimics the economic decision criteria of the operator, but at an aggregate level this procedure is also the source of a large amount of uncertainty. The primary factor that drives operator decision-making, namely, profit, is not incorporated within the current model framework (A4), and other operator-specific conditions are also not considered (A5). To determine the profit of a given structure, the model must account for the expected future values of hydrocarbon price, production, operator cost, reserve estimates, and investment decision-making. Unfortunately, these variables are uncertain, and in most cases, unobservable.

1.4. Conclusions

A statistical description of the explosive removal of offshore structures in the GOM has been presented. The influence of factors such as water depth, planning area, configuration type, structure age, and time upon the application of explosive removals have been examined, and generally-held industry beliefs appear to be "valid," namely, that the application of explosive techniques appear to increase with (i) water depth, (ii) structure complexity, and (iii) age upon removal. Most of the structures that have been removed from the GOM are in the Central planning area (84%) with the remaining removals distributed throughout the Western GOM.

Explosive technology was employed in 954 of the 1,626 structures decommissioned to date – representing in aggregate a 59% explosive removal rate. Caissons were equally likely to be removed with either explosive or nonexplosive methods, while well protector jackets employed explosives 62% of the time and fixed structures were removed with explosives 66% of the time. The applications of explosive methods increase with the complexity of the configuration type, water depth and age of the structure upon removal. The influence of planning area was not a significant factor.

Removal forecasts were developed using a life expectancy and probabilistic removal modeling framework to predict the number of offshore structures that are expected to be removed using explosive technology. The categories employed partition the structure data according to configuration type, water depth, age, and planning area classification, which was subsequently combined with life expectancy and explosive severance modules, to predict removal trends. A description of the modeling process and a summary of the results were outlined. There are significant uncertainties associated with all structure removal forecasts, and forecasting should be viewed relative to the model assumptions and in terms of the "potential" of the likely future impact rather than as a predictive indicator of the actual number of structures expected to be removed.

CHAPTER 2: A BINARY CHOICE SEVERANCE SELECTION MODEL FOR OFFSHORE STRUCTURE REMOVAL

2.1. Introduction

Decommissioning offshore structures is often a severing intensive operation. Cutting is required throughout the structure, above and below the waterline and mudline on braces, pipelines, risers, umbilicals, templates, guideposts, chains, deck equipment and modules. More significant cutting operations are required on elements that are driven into the seafloor, such as multi-string conductors, piling, skirt piling, and stubs which need to be cut 15 feet below the mudline, pulled, and removed from the seabed. Cutting piles and conductors is probably the most critical and important part of a decommissioning project since if the piles and conductors are not cut properly, costly time delays and a potentially dangerous condition can arise during the operation.

A variety of technologies exist to perform severance operations, and the most common cutting methods include abrasive water jet, diamond wire, diver torch, explosive charges, mechanical methods and sand cutters. For severing operations that occur above the waterline, the cutting technique selected is usually dictated by the potential for an explosion. Cold cut methods are used when the potential for an explosion exists; otherwise hot cuts are employed. Cutting in the air zone is conventional, but not hazard-free, since it involves methods which are regularly used for dismantling onshore industrial facilities. Below the waterline, cutting is more specialized. In water depths that do not exceed 150 feet or so, divers perform cuts on simple elements such as braces and pipeline, and for shallow water structures such as caissons, diver torch is sometimes the preferred severance method. In water depths exceeding 150 feet, remotely operated vehicles (ROV's) deployed with abrasive, diamond wire and explosive charges are used for severance operations.

The decision of what cutting method to use will depend on the outcome of a risk-based comparative assessment involving cost, safety, technical, environmental, operational and managerial considerations. To perform a risk-based cost assessment for decommissioning projects *after* the operation has occurred is clearly an imposing (some would say, impossible) task, so we must rely on various proxy variables to estimate the probability that a particular severance technique will be applied in a given situation. The scope of this chapter is motivated by the desire to predict the removal techniques expected to be deployed in the future. Since economic and technical considerations are essentially unobservable, we will rely on a simplified decision model to gain insight on the processes involved in severance selection.

The purpose of this chapter is to describe severance operations within the context of decommissioning and to identify the factors involved in the decision to use explosive/ nonexplosive methods, to quantify the probability that a structure will be removed with explosives, and to establish a relationship between a set of attributes describing a structure and the probability that a particular severance technique will be used.

Background information on the offshore structures and the decommissioning options available to the operator are first described. Additional background information on the various stages of decommissioning and the cutting activities typically performed is presented, along with a summary of the impact of explosive and nonexplosive severance methods. An outline of the regulations associated with the use of explosives is also provided. The factors involved in severance selection are described, the frequency of occurrence of explosive removals is reviewed, and operator use patterns are summarized. Construction of binomial logit and probit models of severance selection conclude the chapter.

2.2. Background Information

2.2.1. Gulf of Mexico Infrastructure: Caissons, well protectors, and fixed platforms were first installed in shallow water, and as development has shifted farther offshore to deeper water, installation types have changed to larger steel structures, tension leg platforms, spars, and subsea completions.

If a reservoir is small or isolated, it will normally be completed with a "minimal" structure – a caisson, well protector, or subsea completion – with flowlines tied back to shore or an accompanying fixed platform. Platforms are classified as major and nonmajor fixed structures. A major structure is defined to include at least two pieces of production equipment or six completions, and will normally include all braced caissons, conventional piled structures with wells, skirt platforms, special platforms, and floating structures (Pulsipher, 1996). Conventional piled platforms without wells (quarters platform, flare pile, storage facility, pipeline junction, metering facilities), single-well caissons, and well protectors are considered nonmajor structures.

Oil and gas structures in the GOM are presented in Table B.1 according to configuration type, water depth, and number of slots available. Major and nonmajor fixed structures comprise slightly more than half of all active GOM structures. Most structures removed to date have been simple structures in shallow waters, and roughly speaking, for every major structure decommissioned, two nonmajor structures have been removed. Over the past decade, the number of structures removed has ranged from a low of 75 to a high of 179, and this range continues to serve as a good indicator on the bounds of expected decommissioning activity in the future.

The most common structure in the GOM is the conventionally piled platform with wells as shown in Figure B.1. In a conventionally piled platform, the platform is pinned to the seabed by long steel tubes called piles which pass through the legs of the structure and act like giant tent pegs. The jacket of the structure provides a protective layer around the conductors, which pass from the seabed up to the topsides, and serve as the conduit to the reservoir (Graff, 1981; McClelland and Reifel, 1986). Piling is sometimes grouted to the jacket leg near the mudline for additional stability and support. In many cases the jacket is installed over one or more exploratory wells with development wells drilled through conductor slots in a central bay. Fixed platforms have been used in the GOM in water depths up to 1,300 feet, but beyond this limit[3] floating production structures are required.

2.2.2. Decommissioning Options and Economic Criteria: Offshore structures normally represent a significant investment and are maintained as long as possible unless economic conditions force their removal. Engineering fatigue, barge collisions, fire, and the occasional

---

[3] Shell's Bullwinkle platform in 1,350 feet water depth stands 1,617 feet tall and is one of the largest fixed structures in the world.

natural disaster may take out a few structures unexpectedly, but for the most part, these factors do not play a significant role in aggregate removal patterns. Structures are designed to last the life of the field.

Abandonment options that are available to the operator include

- Relocation for reuse,
- Removal and scrap, or
- Relocation to an artificial reef site.

The topsides removal and disposal options available in decommissioning projects are shown in Figure B.2 as a decision tree. Oil and gas processing equipment and piping is sent to shore, refurbished and reused, sold for scrap, and/or sent as waste to the landfill. Deck and jacket structures have more options for disposal. The deck and jacket may be scrapped onshore, moved to a new location and reinstalled, or converted to an artificial reef site (Hakam and Thornton, 2000; Thornton, 1989). The complete removal of the jacket is the most frequently used technique in the GOM, occurring in roughly 90% of the total decommissions to date. The remaining 10% of structures that have been decommissioned have been toppled-in-place within an artificial reef or towed to an approved reef site. The Texas and Louisiana artificial reef programs currently maintain over 200 offshore structures throughout the GOM.

The economics of decommissioning are usually considered in terms of "least cost liability" as opposed to "return on investment." Decision criteria associated with abandonment options thus generally favor minimum cost alternatives as the preferred means of most disposals. The factors that determine *when* a structure will be removed, as well as *how* it will be removed, are driven by engineering, economic and safety criteria that is time, location, and operator specific.

2.2.3. Decommissioning is Often a Severing Intensive Operation: The basic aim of a decommissioning project is to render all wells permanently safe and remove most, if not all, surface/seabed signs of production activity. A site should be returned to its "green field" state, but how completely the site should be returned remains a subject of discussion between government, operator, and the public. Cutting operations occur throughout each stage of decommissioning except the first (permitting) and last (site clearance and verification) stage. See Figure B.3. For readers requiring additional information on the activities involved throughout a decommissioning project, the case studies (Hakam and Thornton, 2000; Kirby, 1999; Thornton, 1989) and project checklist (Thornton, 1995) are a good starting point, while more detailed descriptions of the process can be found in (Manago and Williamson, 1998; Pulsipher, 1996; National Research Council, 1985). For a statistical description of GOM activity, see (Kaiser et al., 2002; Thornton and Wiseman, 2000), and for a review of regulatory issues, see (Griffen, 1998; Pulsipher and Daniel, 2000; Shaw, 2000). The phase, timing, and selection of severance operations, and in particular, pile and conductor cutting, is planned to maximize the safety of workers and to minimize the time of the derrick barge on-site. Cutting activities are performed off the critical work path and before the arrival of the barge if the activity can be performed in a cost effective manner.

Well Plugging and Abandonment

A well abandonment program is carried out by injecting cement plugs downhole to seal the wellbore to secure it from future leakage while preserving the remaining natural resources. Techniques used to accomplish this process are based on industry experience, research, and conformance with regulatory standards and requirements (Manago and Williamson, 1998).

A traditional approach begins by "killing" the well with drilling fluids heavy enough to contain any open formation pressures. The Christmas tree is then removed and replaced by a blowout preventer through which the production tubing is removed. Cement is placed across the open perforations and squeezed into the formation to seal off all production intervals and protect aquifers. The production casing is then cut and removed above the top of the cement and a cement plug positioned over the casing stub. The remaining casing strings are then cut and removed close to the surface and a cement plug set across the casing stubs.

Mechanical methods of cutting and sand cutters are primarily associated with well plugging and abandonment (P&A) activities. After wells are plugged and casing tubing cut and pulled, a sand cutter or mechanical cutting tool may be run downhole to cut the conductors, or depending on the preference of the operator/contractor and configuration of the platform, abrasive or explosive severance methods may be applied. In a typical mechanical operation, the tubing and production casing is first cut using a jet cutter – a small explosive blast that utilizes less than five pounds explosive – and then the strings are cut out from $7\frac{5}{8}$ or $13\frac{5}{8}$ inches using a mechanical cutter.

All wellheads and casings are required to be removed to a depth of at least 15 feet below the mudline, or to a depth approved by the District Supervisor. The requirement for removing subsea wellheads or other obstructions may be reduced or eliminated when, in the opinion of the District Supervisor, the wellheads would not constitute a hazard to other users of the seafloor.

Topside Equipment and Deck Preparation

Topside preparation and deck removal is severing intensive. Cold cuts are generally made with pneumatic saws or drills, including diamond wire methods and abrasive techniques. Hot cuts – torch cutting and arc gouging – are used to cut steel when there is no risk of explosion. Arc gouging is used to remove seal welds between steel connections. Burning torches work on the same principle as the arc-gouge, where a burning rod, usually magnesium, is arced with the member to be cut. Diamond wire methods have also been occasionally employed in the GOM to cut the deck from the jacket.

Jacket Preparation

Several severance techniques are used below the waterline. Small cuts made to the jacket bracing and trimming, flowlines, umbilicals, and manifolds are typically performed with divers using burning torches, or if the water depth exceeds the diver capability, ROV's with diver torch or abrasive technology are employed. Intermediate cuts may be required to separate the jacket into vertical sections if the piling extends up through the jacket structure.

Pipeline Abandonment

Federal regulations allow decommissioned OCS pipelines to be left in place when they do not constitute a hazard to navigation, commercial fishing, or other uses of the OCS. Pipelines will generally be removed offshore through the surf zone and capped. Onshore pipeline may be removed completely, or some sections may be abandoned in place if they transition through a sensitive environment. The pipeline end seaward of the surf zone is capped with a steel cap and jetted three feet below the mudline. Most pipelines in the GOM are abandoned in place after cleaning and cutting its structural connections.

The methodology for cutting a pipeline depends on the manner the pipeline is to be recovered. The protective coatings typical of most pipeline sections must first be removed in order to cut the pipe with an arc torch. If a pipeline crosses or is adjacent to an "active" pipeline, chances are it will not be disturbed due to the potential damage that would result if complications arise in the removal. Diamond wire methods, abrasive water jet, and pneumatic saws deployed with diver or ROV are all used to cut pipeline.

Pile and Conductor Severing

Pile and conductor severing is the most critical and typically the most expensive of all the severance operations. Piles are steel tubes welded together and driven through the legs of the jacket and into the seabed to provide stability to the structure, while conductors conduct the oil and gas from the reservoir to the surface. Piles and conductors must be cut and removed a minimum of 15 feet below the mudline. The physical characteristics that describe piles and conductors are important since they determine the technical feasibility of severance options.

Conductors are cut and pulled, if possible, early in the decommissioning process to avoid delay when the barge is on-site. Conductors are configured in various diameters and wall thickness and are characterized by the number of inner casing strings, the location of the strings relative to the conductor (eccentric vs. concentric), and the application of grout within the annuli. Conductors are usually cut with mechanical methods or explosive charges. Grouted annuli are usually easier to cut than annuli with voids since voids dissipate the energy/focus of the abrasive and explosive cutting mechanisms. Eccentricity may also pose a problem for mechanical cutters (Pulsipher, 1996). Mechanical methods are commonly applied to cut conductors during P&A activity, while if conductors are cut when the barge is on-site, then explosive charges will probably be employed.

To sever jacket legs and piles, abrasive cutters and explosive techniques are effective. In principle, mechanical cutting could be used to cut piling, but in practice it is rarely used because piles are only open when a barge is on-site (after removing the deck from the jacket), and with a barge on-site, mechanical cutting is not a cost-effective or efficient way to sever[4]. With a barge on-site, explosives are deployed down the piling and below the mudline, while abrasive cutters can be deployed internally or mounted externally using divers and a track. Obstructions within the pile (such as hangers) will necessitate additional operation or deployment of an external cut. Internal cutting is usually the preferred approach with water jet technology since it does not

---

[4] Redeployment of the barge is usually not an option.

19

require the use of divers to set up the system or jetting operations to access the required mudline depth.

2.2.4. Environmental Consequences of Severance Technology: The use of explosives to cut conductors, well casings, and piles was used for many years without regulation, but in 1986 with the strandings of numerous sea turtles in Texas, concern[5] was raised on the use and application of explosive severance methods. Before 1986, there were no rules or regulations to follow on the use of explosives, and the basic rule of thumb was, "if five pounds does a good job, then ten pounds does a hell of a good job" (DeMarsh, 2000). Since 1986, several regulations have been enacted to help minimize the number of incidental takings[6] and to quantify the impact of using explosives on sea turtles and marine mammals. Observers are currently required for all OCS removal activities using explosive charges >5 lb, and since introduction of the PROP in 1986, only two sea turtles have been killed and three turtles have been injured as a result of explosive severance. The injured sea turtles were rehabilitated and released back into the GOM. Since 1995, more than 750 structures have been removed from federal waters and observers reported no indication of injury or death to bottlenose or spotted dolphins or any other marine mammal (*Federal Register*, 2002b).

The use of explosives to remove offshore platforms does kill and injure fish, but it is important to understand the relative magnitude of the kill relative to other sources of fish mortality in the GOM. For all practical purposes, it is impossible to predict the mortality of fish at a platform during an explosive removal because of the number and uncertain nature of the factors involved. The fish density at the structure, the season of the removal, the range of the fish from the blast, the amount of explosives detonated, and the types of fish are some of the factors that impact fish kill. Between 1993 and 1999, the NOAA Fisheries conducted a study to assess the impact of using explosives at platform removals. Ten platforms from various shallow water locations in the GOM were examined by divers immediately after the explosives were detonated. Total estimated mortality ranged from 2,000-6,000 fish per platform (Gitschlag et al., 2000). The age of the structure, platform size, water depth, water temperature and salinity were not considered factors relevant to fish mortality. Relative to other sources of fish mortality in the GOM (e.g., shrimp trawling), the fish killed by explosive removal are considered negligible. The annual shrimping season in Louisiana and Texas consists of 2, 90-day periods, and it has been estimated that the trawler by-catch is 1,000 fish/trip. If we assume that ½ of the 55,000 registered trawlers in Louisiana and Texas complete 1 trip/day for ¼ of the shrimping season, then the number of fish by-catch on average due to shrimper trawling is roughly 1.25 billion fish per year. On the other hand, if 200 structures are removed from the GOM per year with explosives and the fish kill per

---

[5] The National Marine Fisheries Service (NMFS, currently NOAA Fisheries) sent a letter to the MMS expressing concern regarding stranding events in 1985 and 1986. The stranding events in question coincided with a number of explosive platform removals that were conducted in the state waters of Texas and NMFS suggested that a correlation could exist between the strandings and the use of explosives for decommissioning.

[6] "Take" is defined to mean "to harass, harm, pursue, hunt, shoot, wound, kill, trap, capture, or collect, or to attempt to engage in any such conduct" (16 U.S.C. §1532 (19)). Harass means "an intentional or negligent act or omission which creates the likelihood of injury to wildlife by annoying it to such an extent as to significantly disrupt normal behavior patterns which include, but not limited to breeding, feeding, or sheltering." (50 C.F.R. §17.3) Harm means "an act which actually kills or injures wildlife. Such an act may include significant habitat modification or degradation where it actually kills or injures wildlife by significantly impairing essential behavior patterns, including breeding, feeding, or sheltering." (50 C.F.R. §17.3).

structure is assumed to be an order-of-magnitude greater than the NOAA Fisheries study at 50,000 fish/removal, then the total number of fish kill associated with structure removals is 10 million per year – or less than 1% of the expected shrimper by-catch take.

Nonexplosive cutting methods are considered an ecological and environmentally sensitive severance method since the cutting does not create the impulse and shockwave-induced effects which accompany explosive detonation (Brandon et al., 2000). In mechanical, abrasive water jet, and diamond wire severance technology, a diesel-fueled mechanical motor is employed in the operation which results in vibrations, the emissions of CO_2 and other gases to the atmosphere, and low frequency sound waves into the ocean environment. Abrasive water jet cutting also involves using a fluid and garnet/slag for the cutting mechanism, and so there is the question of the impact of the fluid and garnet on the marine environment. Since the fluid involved in abrasive cutting is water and the garnet is inert, the environmental impact is generally considered inconsequential. Further, the noise level of the supersonic cutting jet is safe for divers and is not considered harmful to marine life. The direct products of nonexplosive cutting processes are water, metal cuttings, and abrasive particles.

There is also an environmental impact associated with the re-suspension of bottom sediments. If the foundation piles are cut below the seabed from the outside, the surrounding sediments will have to be dredged away by suction-dredging or jetted. The use of explosives to cut piling will likely disturb the sediments in the immediate vicinity of the structure. Both operations will cause re-suspension of sediments and contaminants in the cuttings. If the legs/pilings are severed from the inside using abrasive techniques, no significant re-suspension of sediments would ensue. Impacts resulting from re-suspension of bottom sediments include increased water turbidity and mobilization of sediments containing hydrocarbon extraction waste (drill mud, cutting, etc.) in the water column. The magnitude and extent of any turbidity increases would depend on the hydrographic parameters of the area, nature and duration of the activity, and size and composition of the bottom material. The overall impacts to water quality are expected to be temporary in nature and limited in scope to the site (*Federal Register*, 2002a).

2.2.5. The Regulatory Structure of Explosive Severance: To provide adequate protection for marine mammals and sea turtles during OCS decommissionings, explosive-severance activities are subject to regulations promulgated under the OCS Lands Act (OCSLA), ESA, and the Marine Mammal Protection Act (MMPA). In 1988, MMS consulted with NOAA Fisheries on explosive severance and was issued a "generic" or programmatic BO/ITS for sea turtles with terms and conditions that are currently still in effect (see http://www.gomr.mms.gov/homepg/regulate/environ/generic-consultation.pdf). In 1989, the American Petroleum Institute (API) petitioned NOAA Fisheries under Subpart I of the MMPA for the incidental take of spotted and bottlenose dolphins during removal operations (both explosive and nonexplosive severance). NOAA Fisheries promulgated incidental-take regulations under Subpart M of the MMPA regulations in October 1995 with mitigative measures similar to those listed in the 1988, ESA BO/ITS. Subpart M expired in November 2000, after which, NOAA Fisheries released an Interim Subpart M in August 2002, and since its expiration on February 2, 2004, operators have continued to follow the similar ESA BO/ITS terms and conditions for both turtles and marine mammals.

The MMS has a Notice to Lessees and Operators (NTL), No. 24-G06 (http://www.gomr.mms.gov/homepg/regulate/regs/ntls/ntl04-g06.html), that summarizes all of the current regulations and conditions for explosive-severance activities. For additional background information on regulations concerning endangered species, see McKay et al., (2002) and Roberts and Hollingshead, (2000). A flowchart of the procedures involved with explosive cutting is shown in Figure B.4.

- Qualified observer(s), as approved by the NOAA Fisheries, must be used to monitor the area around the site for a period of 48 hours prior to, during, and after the detonation of explosives.

- If sea turtles are observed in the area and are thought to be resident[7] at the site, pre-detonation and post-detonation diver surveys must be conducted.

- On day(s) that blasting operations occur, a 30-minute aerial survey must be conducted one hour before and one hour after each blasting episode. To ensure that no marine mammals are within the designed 1,000 yards safety zone nor are likely to enter the designated safety zone prior to or at the time of detonation, the pre-detonation survey must encompass all waters within one nautical mile of the structure.

- If sea turtles and/or marine mammals are observed within 1,000 yards of the structure prior to detonation, blasting must be delayed. The delay must remain in affect until the sea turtles and/or marine mammals are beyond 1,000 yards of the platform. The aerial survey must be repeated prior to resuming detonation of the charges.

- If sea turtles and/or marine mammals are observed within 1,000 yards of the structure an additional survey must be performed, involving either a diver survey dedicated to marine mammals and sea turtles within 24 hours after the detonation event or an aerial/vessel survey within 2-7 days after the blast. The aerial/vessel survey must concentrate down-current from the structure.

- The NOAA Fisheries-approved observer may waive post-detonation monitoring provided no marine mammal was sighted during either the aerial surveys before detonation or during the 48 hour pre-detonation observer monitoring period.

- Explosives can be detonated no sooner that 1-hour after sunrise and no later than 1 hour before sunset. Special circumstances may allow for modification of these times if permitted by the NOAA Fisheries personnel on-site.

- During all diving operations, divers are required to scan the area around the platform for sea turtles and marine mammals. Any sightings must be reported to the NOAA Fisheries personnel of the agent of the holder of the Letter of Authorization upon surfacing.

- In water depth of 150 feet or greater, or in cases where divers are not deployed in the course of normal removal operations, a remotely operated vehicle must be deployed prior to detonation to scan areas below structures. If marine mammals are sighted, the

---

[7] The majority of turtle sightings are turtles "in transit". In cases where turtles are believed to be "resident" to the structure, either the turtles are relocated or the severance operation is performed with nonexplosive methods.

ROV operator must inform either the NOAA Fisheries observer or the agent of the holder of the Letter of Authorization immediately.

- In water depth of 328 feet (100 meters) or greater, passive acoustic detection must be employed prior to detonation. If marine mammals are detected by the acoustic device, the operator must inform either the NOAA Fisheries observer or the agent of the holder of the Letter of Authorization.

- After the explosives are detonated, if sea turtles and/or marine mammals are sighted, either dead or injured, attempts should be made to recover the animals.

- A report summarizing the results of structure removal activities, mitigation measures, monitoring efforts, and other information as required by a Letter of Authorization, must be submitted to the NOAA Fisheries Regional Administrator within 30 days of the removal of the structure.

Regulations intended to protect sea turtles and marine mammals provide built-in incentives for using nonexplosive techniques. The time, scheduling, and expense of coordinating the NOAA Fisheries observers during explosive removals, the restrictions on using explosives, and the possible delays associated with the presence of marine life, encourages operators to consider alternative severance methods – when alternative methods are feasible.

2.3. Factors Involved in Severance Selection

A large number of factors are potentially involved in selecting the severance technique for a specific job, with cost, safety, risk of failure, and technical feasibility the primary factors that are considered when alternative options are available. Many different severance operations are required during decommissioning, and depending upon the job, more than one alternative is usually available (Figure B.3). In general, cutting techniques are expected to be reliable, flexible, adaptable, safe, cost effective and environmentally sensitive (National Research Council, 1986). If a cutting technique fails with respect to one or more of these factors, or if an operator has more than one "bad experience" with a particular method, then chances are that the technology will not gain in popularity or acceptance among GOM contractors.

Variables that drive the cost and risk associated with a specific severance technique are numerous and involve factors such as the location and nature of the site, sensitivity of the marine habitat, structural characteristics, the amount of pre-planning involved, salvage/reuse decisions of the operator, marine equipment availability, operator experience and preference, contractor experience and preference, the number of jobs the contractor is scheduled to perform and the schedule of the operation, and market conditions. Some of these variables are observable, but the degree of correlation between the observable variables and severance decision factors is expected to be weak, and so the extent to which cutting methods can be accurately predicted based on these factors remains uncertain.

2.3.1. Direct Cost: The cost of a derrick barge is at least an order-of-magnitude larger than the cost of a cutting spread, and so cutting decisions are *not* expected to be a primary focus in decommissioning operations *unless* they negatively impact the time on-site of a derrick barge. The direct cost of a cutting spread is at most $10,000/day (Kaiser and Pulsipher, 2003), and when compared to a derrick barge spread of \$100,000-\$300,000/day, it is clear that cutting techniques will not drive decommissioning activities. The cost to sever piles and conductors is generally less than 1-3% of the total cost to decommission the structure.

2.3.2. Cost of Failure: If the cutting operation is not successful on the first attempt, then the operator will assume the cost of failure and the additional time required to re-shoot or re-cut the tubular element(s). In Figure B.5 the abrasive cutting process is charted. Contractors typically charge at work rates that depend upon the critical[8] path crane vessel time. Normally, if "extra work" is required that alters the critical path, the contractor charges the operator rates for equipment and personnel affected. If extra work is required that does not alter the critical path crane vessel time, the operator is charged a different (substantially smaller) hourly composite rate. The cost of a failed cut thus depends on the timing of the cut relative to the operational activity of the barge. There is a significant difference between a crew *preparing* for a cutting operation, while other barge activities are on-going, versus a crew *cutting* while other barge activities wait for the operation to finish. Failure to cut a conductor off the critical path is not nearly as significant as a cutting failure that delays barge activity. The expected cost of failure is a primary decision factor in selecting a cutting method (Greca, 1996).

2.3.3. Human Safety: The offshore environment is a potentially hazardous location which presents special risk to the personnel involved in the operations. At each stage of decommissioning there is the potential for work injury and fatality. Cutting and welding steel tubes, setting up rigging, diving, moving cranes, hydraulic equipment, explosive charges, and old rusty structures create the potential for a hazardous work environment, and so proper precautions are required to ensure that operations are performed in as safe a manner as possible. Fortunately, decommissioning activities are fairly standard, of short duration, and relatively safe if properly planned and executed. A typical decommissioning operation will take anywhere from 7-14 days to complete.

Cutting the piles and conductors is probably the most critical and important part of a decommissioning project since if the piles and conductors are not cut properly, a potentially dangerous condition could arise during the lift. The bottom cuts on anchor piles and conductors must be "clean" and "complete" to allow for a safe operation. Incomplete cuts pose a serious danger to the stability of the vessel during lift.

2.3.4. Environmental Issues and Structure Disposition: Under some circumstances, the choice of severance method may be determined exclusively by the location of the structure and/or the decision to re-use the jacket; e.g.,

- Structure is located in a known turtle habitat,

- Structure is located in an artificial reef planning area, or

- Structure jacket will be re-used.

---

[8] Critical path activities are considered "bottleneck" operations that "use up" barge time and are charged at a premium.

These circumstances do not occur frequently – probably in about 10-15% of the structures removed from the GOM – but they do occur (e.g., see Ness et al., 1996; O'Connor, 1998) .

If the jacket is to be re-used or the structure is located in a known turtle habitat, then nonexplosive methods will likely be used if technically feasible. Clean cuts are desirable to avoid the diver cost/risk associated with flared piles and the possible damage that can occur to the re-used jacket with explosive cutting. If a structure is located in an artificial reef planning area and it can be toppled-in-place, then the piles and conductors are severed and the jacket is pushed over to form the reef (Dauterive, 2001; Reggio, 1989). If the structure does not satisfy the minimum 85 feet waterline clearance, then the structure will need to be cut in the water column and partially removed, that is, the top of the re-used jacket will be cut and placed on its side near the bottom of the jacket which will be left in place. In a partial removal, the piles do not need to be severed from the bottom structure, and since the use of explosives is prohibited in the water column, abrasive water jet, diver torch, or diamond wire methods are used to make the mid-water cuts.

2.3.5. Operator Experience and Preference: The project management team overseeing the decommissioning activities, in consultation with the operator, prepares the bid package and specifies the work requirements to be performed. This information will include special requests, such as platform and jacket disposition, and preference (if any) for the severance method to be employed. The operator may also have special concerns or preferences that dictate that a specific method be employed. For example, between November 15, 2000 – August 1, 2002 some operators (e.g., Shell, El Paso) specifically requested that contractors use nonexplosive methods for cutting since the expiration of Subpart M, MMPA took away NOAA Fisheries authority to issue Letters of Authorization for removal activities (Guegel, 2001). As a result of Subpart M's expiration, operators were potentially exposed to NOAA Fisheries-assessed penalties and fines should a bottlenose or spotted dolphin take be recorded during a structure removal activity (either explosive- or nonexplosive severance).

There are also many other reasons for operator preference. Some operators may consider the benefits offered by explosives to be outweighted by public perception, while other operators – especially operators who have never decommissioned a structure – may want the first removal to use a "standard" (explosive) cutting method to avoid complications. All operators want to avoid cost overruns and minimize potential risk to offshore personnel. Unfortunately, unless preferences are known or readily identifiable, they are considered unobservable and must be accounted for indirectly through proxy variables.

2.3.6. Operational Scheduling: Conductor severing and recovery may be completed as part of well plugging and abandonment activities unless the platform configuration, equipment availability or scheduling of the activities prevent the operation. Conductors are cut and pulled, if possible, early in the decommissioning process to avoid delay when the barge is on-site. Mechanical casing cutters and sand cutters can be used to perform the cut if a crane is available on the platform for the deployment of the tool. To verify a complete cut, a jacking spread may be used to lift the conductor after the severing attempt. To jack the conductors (prove the cut), the platform must have the structural capacity to provide a point to jack against and have a crane large enough to set the cutter, jacks, and load spreading beams. If tubing and casing strings have not been cut prior to the arrival of the derrick barge, then explosive charges will likely be used to cut all the elements at once. Mechanical and/or sand cutters are rarely deployed with a derrick barge on-site due to the time-consuming and inefficient nature of the operation.

2.3.7. Contractor Experience and Preference: If the contractor has several removals to make, then the preference is to cut as quickly and as safely as possible subject to the technological and operational requirements of the job. If explosives are required on one structural element, then a preference may arise to blow all the elements at once rather than "mix" explosive and nonexplosive severance methods, and as mentioned earlier, if pre-cuts are not performed on the conductors, then explosives are more likely to be employed to sever all the elements when the barge is on-site. On a few decommissioning projects, abrasive water jet and explosive cutting crews have served in a contingency role, but since back-up crews add significantly to the cost of the service, cutting redundancy is not standard practice.

2.3.8. Structure Characteristics: Pile and conductor severing is the most critical and typically the most expensive of all the severance operations required on the structure. The physical characteristics that describe piles and conductors are important since they allow engineers to determine the technical feasibility and potential problems of removal options.

Conductors are configured in various diameters and wall thicknesses and are characterized by the number of inner casing strings, the location of the strings relative to the conductor (eccentric vs. concentric), and whether or not the annuli are grouted. Conductors typically contain multiple strings of casing, eccentric within the well, and grouted with voids. Grouted annuli are usually easier to cut than annuli with voids since voids generally dissipate the energy and focus of the abrasive cutting mechanisms. The preferred way to cut conductors is with mechanical methods or explosives charges, while piles are effectively cut with abrasive methods and explosive charges. Since piling cannot be examined before the topsides are lifted off the jacket, bulk explosives are usually preferred for piling since they can be sized for unexpected field conditions and give a clear indication of a complete cut.

2.3.9. Structure Age: Contractors select severance methods that are cost effective, reliable, efficient, adaptable, safe and environmentally sensitive. If a structure is old, it is less likely to have accurate records and drawings available, and if accurate information is not available to the cutting crew before the cut is performed, the ability to plan and anticipate potential problems in the operation is constrained. Old structures are also less likely to be re-used, and so we would expect explosive methods to be more frequently applied as the age of the structure increases.

2.3.10. Cutting Reliability: The ability of a severance technique to perform a cut, and to perform it reliably, is a significant factor and one of the most important criteria contractors consider in their selection of a removal method – especially if cutting is to occur near critical path barge activities. No cutting technique is 100% reliable and so the operator/contractor's experience with the technology and their perception of reliability is as important as aggregate reliability statistics. Unfortunately, it is difficult to acquire reliability statistics that can be meaningfully compared across severance methods due to the lack of appropriate data. Industry consensus is that explosives remain the most predictable, flexible, and reliable severance technique (Kaiser and Pulsipher, 2003; Pulsipher et al., 1996).

2.3.11. Configuration Type: Nonexplosive methods usually carry less financial and operational risk with shallow water, simple structures than for complex, deep water structures (National Research Council, 1986). Mechanical and sand cutters have been used effectively on shallow water caissons and small well protector jackets, and large caissons have been effectively cut by divers. As the complexity, size, and water depth of a structure increases, however, the reliability of nonexplosive methods decreases while the cost and risk/uncertainty of operations tend to increase. On large platforms, especially platforms with wells, the preferred severance method is with explosives. There is not a "smooth" transition that occurs as a function of water depth or structure complexity, but generally speaking, we would suspect that as the complexity and water depth of a structure increases, explosive methods should be applied more frequently, and this is borne out by statistical analysis of the removal data. Explosives cut quickly and reliably and crew exposure time is minimal. For special structures such as skirt-piled[9] platforms, mechanical, abrasive, and diver cuts are usually not feasible and the tubular elements are generally stabbed with explosives using an ROV.

2.4. The Probability of an Explosive Removal

The choice of which severance technique is used to cut the piles and conductors of a structure depends primarily on factors which are unobservable or uncertain, and so it is clear that for a given structure we can only seek the probability that explosive methods will be applied. It is necessary to proxy the unobservable variables with factors that are accessible and are "reasonably" reflective of the offshore environment. Configuration type, water depth, and the structure age upon removal are public information, while the characteristics of the piles and conductors associated with each structure; e.g., number, size, application of grout, number of casing strings, etc. are typically not recorded. It would be preferable to perform the analysis at the lowest possible level of aggregation – which in this case is described through the characteristics of the structure – and then to "work up" through various aggregation strategies. Due to deficiencies in the MMS database, it was not possible to explore the impact of aggregation schemes on the probability measures, but for practical purposes due to the nature of the problem, additional data would probably contribute only marginally to increased understanding.

The manner in which data is reported also needs to be interpreted carefully since an "explosive removal" suggests that *all* the piles and conductors of a structure are explosively severed, while in fact as we have previously discussed, various severance techniques may be used throughout the structure to cut the tubular members. The diversity of the severance procedures are not captured through the MMS database, and this "information loss" restricts the degree (and ultimately limits the ability) to which a decision model can be constructed. The percentage values computed thus need to be interpreted as the number of structures in which explosives were used *at least once* during decommissioning. It would be better to compute the frequency of application of explosive severance as a function of the number and size and type of tubular elements that are cut, but due to deficiencies in data collection, this characterization is not possible.

---

[9] Skirt-piled platforms are predominately used in deepwater with skirt piling driven through sleeves to provide additional axial and lateral load bearing support.

Caissons are the most likely to be removed using nonexplosive methods, and well protectors and fixed platforms, if removed with nonexplosive technology, is more commonly performed in shallow water (Table B.2). As water depth increases, the chance of using explosives increases slightly across all configuration types. Refined partitions of the water depth data (e.g., using 3 meter, 10 meter, and 25 meter increments) indicated no observable "trends," and so the consideration of water depth as a relevant factor is questionable. The percentage of structures removed using explosive techniques is depicted in Table B.3 according to age upon removal, configuration type, and water depth. The use of nonexplosive methods is most common across all configuration types within the 0-10 year category when the structure has the greatest chance for re-use, and as the age and water depth of structures increase, roughly speaking, the probability of an explosive removal also increases.

2.5. Operator Practice in the Gulf of Mexico

Since 1986, 1,626 structures operated by 127 companies have been removed in the GOM. A few hundred structures were removed before this time, but the use of explosives for decommissioning was not documented formally by operators or government agencies. Twelve of the 127 companies are responsible for half of all structures removed, while the "top 36" companies, each removing at least eleven structures, account for 80% of all abandonments (refer to Table B.4). Companies that have removed ten structures or less comprise the "bottom 91" category and contribute the remaining 20% of decommissioned structures. Summary statistics present a complicated picture of operator behavior.

Chevron (now ChevronTexaco) has removed the most structures in the GOM as shown in Figure B.6, and when combined with Texaco's structures, comprise roughly 20% of all removals. Company practice in the application of explosives is diverse and centered around a mean explosive removal rate of 63% (top 12) and 57% (middle 24).

Companies that have removed ten structures or less are more likely to use *only* explosive technology. Over one-third of the "bottom 91" operators used explosives exclusively, compared to only one company out of the "top 36" operators. The "top 36" operators use explosives for removal on average 58% of the time compared with a 51% application rate for the "bottom 91" companies. The primary statistical difference between the two groups is that the standard deviation for the "bottom 91" companies is approximately equal to its mean and twice as large as the "top 36." In other words, there is a wide degree of variability in the application of explosive technology among the major players, but among the "bottom 91" firms, the variability is even greater (see Figure B.7).

The percentage measures depicted in Figure B.6 and Figure B.7 are *not* an especially good indicator to compare the use of explosives across operator since structural characteristics are not considered in the analysis. Normalization of the data can be performed on a company-wide basis, but the results are not expected to provide any additional useful information.

2.6. Binomial Logit and Probit Models of Severance Selection

2.6.1. Model Development: A binary-choice severance selection model assumes that the operator is faced with a choice between two alternatives (explosive versus nonexplosive severance) and that the choice of which cutting method to select depends on characteristics that are identifiable. The requirements of the binary-choice model are quite strong, since as we have described previously, many important characteristics of the severance decision are not observable, and hence, not possible to incorporate within a model. It is nonetheless useful to explore the use of an econometric model since it quantifies the probability of an explosive cut and provides additional insight into the data interpretation.

The purpose of a qualitative choice model is to determine the probability that a structure with a given set of attributes will realize a specific removal method. We establish a relationship between a set of attributes describing a structure and the probability that the operator will make a given choice. A binomial logit and probit model is constructed to model the probability that a structure will be explosively severed. Define the dummy variable

$$D = \begin{cases} 1, & \text{structure is removed with explosives} \\ 0, & \text{otherwise.} \end{cases}$$

If we collect a sample of structures that have been removed from the GOM, it is clear that the outcome D is a random variable that will only be known after the sample is drawn.

There are many relevant variables that impact the selection of the severance method, and some of the observable characteristics include the configuration type, ST; age upon removal, AGE; and water depth of the structure, WD. The general model is thus written

$$D = f(ST, AGE, WD),$$

where,

$$ST = \begin{cases} 1, & \text{if the structure is a caisson} \\ 2, & \text{if the structure is a well protector} \\ 3, & \text{if the structure is a fixed platform,} \end{cases}$$

AGE = Age of the structure upon removal,

WD = Water depth of the structure.

The logit model is based on the use of a cumulative logistic probability function which is specified as

$$F(l) = P(L \le l) = \frac{1}{1 + e^{-l}},$$

while the probit model is associated with the cumulative normal probability function which is written as

$$F(z) = P(Z \leq z) = \int_{-\infty}^{z} \frac{1}{\sqrt{2\pi}} e^{-5u^2} du .$$

If the probability of an explosive removal is related to the variables in a linear fashion, such as

$$E(D) = \beta_0 + \beta_1 ST + \beta_2 AGE + \beta_3 WD + \varepsilon,$$

then the probability that the observed value D takes the value 1 in the logit model is given by

$$P(L \leq \beta_0 + \beta_1 ST + \beta_2 AGE + \beta_3 WD) = F(\beta_0 + \beta_1 ST + \beta_2 AGE + \beta_3 WD)$$

$$= \frac{1}{1 + e^{-(\beta_0 + \beta_1 ST + \beta_2 AGE + \beta_3 WD)}},$$

while the probit model expresses the probability as

$$P(Z \leq \beta_0 + \beta_1 ST + \beta_2 AGE + \beta_3 WD) = F(\beta_0 + \beta_1 ST + \beta_2 AGE + \beta_3 WD).$$

By construction, the value of the probability in the logit and probit models will lie in the interval (0, 1) and represent the conditional probability that an event occurs given values of *ST*, *AGE*, and *WD*.

2.6.2. Model Results: The expected signs for the coefficients of the regression model are straightforward, since as the complexity, age, and water depth of a structure increases, so does the probability that explosives will be used (recall Tables B.2 and B.3). Hence, the coefficients of the regression models are expected to be positive.

The sample set is the universe of structures removed in the GOM between 1986-2002 and consists of over 1,500 individual observations. The coefficients of the logit and probit model are estimated using maximum likelihood, an iterative estimation technique useful for nonlinear equations (Berndt, 1991).

The results of the estimation are shown in Table B.5. The estimated coefficients have the expected signs and two coefficients are significantly different from zero. The water depth does not appear to be a relevant factor. The logit and probit model results are quite similar since they are both based on cumulative distribution functions. The value of R^2 cannot be relied on as a measure of the overall fit of the model with a dummy dependent variable, but one alternative is to compute R_p^2, the percentage of the observations that the estimated equation correctly explains. To use this approach, compute

$$R_p^2 = \frac{\text{number of observations "predicted" correctly}}{\text{total number of observations}}.$$

30

R_p^2 is not used universally, but it is a convenient and easily interpreted measure (Studenmund, 2001). The R_p^2 indicates that the equation correctly "predicted" over 60% of the sample based on three variables – only slightly better than guessing.

2.6.3. Interpretation and Application of the Probit and Logit Model: The effect of a unit change in factor *ST* holding the factors *AGE* and *WD* fixed, on the probability that *D* = 1, is given by

$$\frac{dP}{d(ST)} = \frac{dF(t)}{dt} \cdot \frac{dt}{d(ST)} = f(\beta_0 + \beta_1 ST + \beta_2 AGE + \beta_3 WD)\beta_1.$$

To predict the probability that a structure will be removed with explosives, we use the probability model $P = F(\beta_0 + \beta_1 ST + \beta_2 AGE + \beta_3 WD)$. If we obtain estimates a_0, a_1, a_2, and a_3 of the unknown parameters, we then estimate the probability to be

$$\hat{P} = F(a_0 + a_1 ST + a_2 AGE + a_3 WD).$$

By comparing the probability to a threshold value, the choice of severance method is predicted using one-half as the cutoff:

$$\hat{D} = \begin{cases} 1, \hat{P} > 0.5 \\ 0, \hat{P} \leq 0.5. \end{cases}$$

Example

Consider an 8-year old well protector in the WGOM in 50 feet water depth. The probability the structure will be removed with explosives is estimated to be $\hat{P} = 0.61$.

2.6.4. Limitations of the Analysis: The binomial logit and probit models are limited by several conditions:

- Ability to capture the relevant factors involved in decision making,
- Ability to adequately model the identified factors, and
- Ability to extract sufficient data to support the modeling effort.

Each of these issues plays a role in the construction and development of the qualitative choice model.

The large number of factors potentially involved in the selection of a severance method provides the first indication that the decisions involved in severance are complicated and difficult to model. The proliferation of factors is the first clue that simple models can not provide a complete reflection of the decision framework, and in fact, it is unlikely that *any* model of the decision process can incorporate all the relevant factors and most certainly a simple model will not do the trick. The modeling process in this case is only useful to quantify the data in a more sophisticated manner. The model does not reduce or eliminate uncertainty or provide additional information that is not already captured through probability tables. Relevant company and site specific information (e.g., equipment available at the time of the removal, the amount of pre-planning involved in the removal, the contractors preference and the operational scheduling, the terms of the contract, the quality of the structure blueprints, etc.) can play an important role in the choice of removal method, but because these factors are unobservable, they cannot be statistically analyzed. It is thus clear that a significant portion of the decision making framework cannot be incorporated within the model. The relationships established should thus be viewed as interpretative rather than as causal in nature.

The MMS tracks the number of structures removed, the manner of severance, and the structure classification, and this data provides the basis for the model construction. The characteristics of the structure, including the number and size of the tubular members, the application of grout, and the manner of removal of each tubular element do not form part of the MMS data set, and thus also cannot be incorporated within the decision model. It is unlikely that the inclusion of more refined data at a lower level of aggregation will provide useful information, however, and so in principle, the limitations of the MMS database are not effectual.

2.7. Conclusions

Decommissioning activities in the GOM are driven by economics and technological requirements and governed by federal regulation. Decisions about when and how a structure is decommissioned involve complicated issues of environmental protection, safety, cost, and strategic opportunity, and the factors that influence the timing of removal as well as the manner in which a structure is removed are complicated and depend as much on the technical requirements and cost as on the preferences established by the operator and/or project management team overseeing the decommissioning project.

The purpose of this chapter is to describe the issues involved in severance selection and the factors involved in the decision to use explosive/nonexplosive severance methods, to quantify the probability that a structure will be removed with explosives, and to establish a relationship between a set of attributes describing a structure and the probability that the operator will select a given severance technique. Models were described to help explain the choice of severance technique through a closed-form equation, but because such models are not directly linked to structure characteristics, the veracity of the relationships described is purely statistical in nature and should not be considered causal.

CHAPTER 3: MODELING THE DECOMMISSIONING TIME OF OFFSHORE STRUCTURES

3.1. Introduction

Business decisions accompany every stage of oil and gas exploration and production. A company acquires a lease or contract area based on geological and geophysical data and conceptual plays, and then invests in additional data and manpower to refine their knowledge of the region. If the results of the analysis are encouraging, exploratory drilling may result. If drilling is successful (and most often it is not), the company will confirm and delineate the field, and if the field is judged to be economic, the company will develop and produce the reserves in accord with its risk-reward perceptions of development in the area. Enhanced recovery projects may be added during the field's producing life if the incremental economics are positive. Frequently, operators will divest their property or form a joint venture/farmout type arrangement before the economic limit is reached. When the production revenue of the structure equals the operating costs, abandonment follows.

At any point in time during the life cycle of a field, and depending upon the prevailing and expected future economics, technologic development, strategic objectives, political trends, and contract terms, the operator has to make short-term operational and long-term strategic planning decisions. Four primary options exist:

- Produce. Hold the asset, produce, and manage the declining reserves.

- Invest. Invest in the asset to maintain or increase production.

- Divest. Sell all or a portion of the working interest ownership.

- Decommission. Stop production and remove the asset in accord with regulatory requirements.

Produce

Early in the life of a field after the development wells have been drilled, the field is produced according to equipment capacity and operating constraints. Capital expenditures decline quickly after development is complete, and after the field begins to flow, gross revenues turn positive. Once the exploration and development costs of the investment have been borne, the variable cost of production is usually fairly small, and the operator needs only to produce to achieve cash flow. The cumulative net cash flow breaks even at payout, after which the cash flow remains positive until such time that additional capital investments are required.

Invest

Investment will alter the production profile and will typically extend the life of the asset. If a field requires major new investment such as significant workovers or the introduction of secondary techniques to maintain production, then the field is likely to be considered a candidate for divestiture or abandonment. Major and large independent operators frequently divest property before the economic limit is reached if the rate of return does not meet a minimum threshold or the strategic goals of the company change; e.g., the operator may redefine their core assets or need to raise capital to pursue frontier development. This may lead to the removal of the structure, or if the field can still be operated profitably, then it may be purchased and operated by another firm.

Divest

Property divestment is a key feature of offshore operations. Operators regularly "carve up" assets and sell or subject them to various joint venture/farmout type arrangements throughout the life cycle of the field. This is sometimes referred to colorfully as an asset "moving down the food chain," and in most instances, properties change hands three or more times before the structure is finally decommissioned. Companies buy producing properties and then implement a comprehensive program to increase production, typically involving drilling new stepout or infill wells and recompleting existing wells. Companies specializing in marginal production focus on operating mature fields in a geographic region where they already own infrastructure. The fact that niche operators can manage marginal assets at a profit is due in part to their lower overheads, lower expected rate of return, scale economies, and other strategic factors; e.g., the operator may be a subsidiary of a construction company which serves as a feeder to the parent. Divestment frequently acts to extend lease life, recover greater quantities of hydrocarbons, and ultimately, delay the "expected" decommissioning date of the structure.

Decommission

Decommissioning represents a liability as opposed to an investment, and so the pressure for an operator to decommission a structure is not nearly as strongly driven as installation activities. Delaying decommissioning has an economic value for the firm since it defers expenditure while allowing the deferred funds to be invested in productive (profit-generating) activities (Roberts and Mitchell, 1997). Operators may maintain uneconomic production to see if new extraction technology can increase field life or hold (requalify) infrastructure for new (or marginal) development plans. Delaying decommissioning can also play a beneficial role in the extraction of natural resources in terms of economic efficiency and optimal resource management. Hydrocarbon infrastructure represents a social investment, but once wells are plugged and abandoned and infrastructure removed, hydrocarbon resources "left behind" in the field or on nearby acreage is either "lost" or only available in the future at a higher cost to society. Various proposals have been suggested over the years to delay structure decommissioning to achieve better economic efficiency; e.g., allow the operator to pay rent to the government to maintain the structure in place after production ceases, but because of liability, maintenance, and environmental concerns, these ideas have never received much support.

The purpose of this chapter is to develop the analytic framework to predict the decommissioning time of an offshore structure. The basic economic model and the parameters required to drive the models are discussed. The GOM is used as the reference case in discussion, but the framework is generally applicable to other offshore regions in the world, when adjusted for the governing fiscal regime and legislative requirements. Four models of decommissioning are developed, ranging from a simple, production-based forecast to a risked, net present value approach. A meta-modeling methodology is employed to analyze the simulation results, and a detailed example is used to illustrate the approach. The limitations of the analysis are described and conclusions complete the chapter.

3.2. After-Tax Net Cash Flow Analysis

3.2.1. Units of Analysis: Four units of analysis are typically employed in hydrocarbon modeling: well, structure, lease, and field. The unit of categorization employed depends upon the requirements of the problem and data availability. Production problems are examined at the wellhead, while operators consider development planning and cost allocation on a lease or field basis. The U.S. government requires royalty, rent, and bonus bid payment to be paid on a lease basis.

Holes must be drilled into the Earth to search for and produce oil and gas. These holes, or wells, produce a mixture of oil, gas, water, and other materials which must be separated and treated prior to its transport to market.

A well produces from a reservoir – a porous, permeable rock body, sort of a sponge – lying underneath an impervious layer of rock that traps the resource. Several reservoirs located within a "common" geologic feature are called a field and can consist of a single reservoir or multiple reservoirs. The pressure on the fluid in a reservoir rock causes the fluids to flow through the pores into the well. The reservoir drive comes from fluid expansion, rock expansion, and/or gravity. There are four basic types of reservoir drives for oil reservoirs: 1) dissolved gas drive, 2) free-gas cap expansion drive, 3) water drive, and 4) gravity. Every oil reservoir has at least one, and sometimes two, of these reservoir drives. Gas reservoirs have either an expansion-gas or water drive (Hyne, 1995).

Each well is associated with a structure which is identified by its leasehold and type. Offshore structures vary significantly depending on the productivity of the reservoir and the quality of the produced hydrocarbons; logistical considerations in moving production to market; and the lead time required to acquire or design and construct platforms, drilling rigs, production equipment, and other downstream facilities. The basic size and function of an offshore structure result from the requirements of the development plan (Gerwich, 2000). Typically, several wells are associated with a structure, and more than one structure is located on a lease.

Lease terms and dimensions vary with the time of the auction and the location of the lease, but most give the leaseholder the exclusive right to explore for oil and gas for a period of 5-10 years. The terms of the lease extend for as long as the lease is productive or development/drilling activities are progressing.

The amount of hydrocarbons produced by well w_i in year t is denoted by $Q(w_i, t)$. Production is expressed separately in terms of barrels (bbl) of oil or cubic feet (cf) of gas, or in terms of a single stream as barrels of oil equivalent (BOE). Barrels of oil equivalent are the amount of natural gas that has the same heat content of an average barrel of oil[10]. The annual hydrocarbon production associated with structure s_i is the aggregate of its collection of wells, $\{w_1, \ldots, w_{n_i}\}$:

$$Q(s_i, t) = \sum_{j=1}^{n_i} Q(w_j, t).$$

Similarly, the hydrocarbon production on lease l at time t is denoted by $Q(l, t)$, and is determined as the collection of all the structures contained on the lease, $\{s_1, \ldots, s_m\}$:

$$Q(l, t) = \sum_{i=1}^{m} Q(s_i, t).$$

3.2.2. After-Tax Net Cash Flow: The net cash flow vector of an investment is the cash received less the cash spent during a given period, usually taken as one year, over the life of the project. Using structure s as the basic unit of analysis, the after-tax net cash flow in year t is computed as

$$NCF(s,t) = GR(s,t) - ROY(s,t) - CAPEX(s,t) - OPEX(s,t) - TAX(s,t) - OTHER(s,t),$$

where,

$NCF(s,t)$ = After-tax net cash flow of structure s in year t,

$GR(s,t)$ = Gross revenues of structure s in year t,

$ROY(s,t)$ = Total royalties paid by structure s in year t,

$CAPEX(s,t)$ = Total capital expenditures of structure s in year t,

$OPEX(s,t)$ = Total operating expenditures of structure s in year t,

$TAX(s,t)$ = Total taxes paid by structure s in year t,

$OTHER(s,t)$ = Other expenditures of structure s in year t.

3.2.3. Cash Flow Components: The gross revenues in year t due to the sale of hydrocarbons is defined as

$$GR(s,t) = g^o(s,t)\, P^o(s,t)\, Q^o(s,t) + g^g(s,t) P^g(s,t)\, Q^g(s,t),$$

where,

$g^o(s,t), g^g(s,t)$ = Conversion factor of oil (o), gas (g) in year t,

$P^o(s,t), P^g(s,t)$ = Average oil, gas benchmark price in year t,

$Q^o(s,t), Q^g(s,t)$ = Total oil, gas production in year t.

---

[10] One BOE is about 6 Mcf of gas.

There are four basic types of hydrocarbon molecules, called the hydrocarbon series, in each crude oil: paraffins, naphthenes, aromatics, and asphaltics. The relative percentage of each series molecule controls the chemical and physical properties of the oil. Natural gas is composed of hydrocarbon molecules ranging from one to four carbon atoms in length: methane (CH_4), ethane (C_2H_6), propane (C_3H_8), and butane (C_4H_{10}). The conversion factor (or "quality" of the production stream) depends on the physical characteristics of the hydrocarbons and is a function[11] of the API gravity, the sulfur content and the gas-oil ratio (GOR).

API Gravity

The API gravity of crude oil is a measure of the density or weight of the oil. Average crude has a 25° to 35° range, with light oils falling between 35° to 45° and heavy oils below 25°. Light crude receives a higher price relative to heavy crudes because they tend to have more gasoline by volume.

Sulfur Content

The sulfur content for most crude oils falls between 1% and 2.5%, with 1% sulfur content considered "sweet" crude and 2.5% sulfur considered "sour." Sweet crude is priced at a premium relative to sour crude. Hydrogen sulfide can occur either mixed with natural gas or by itself. Hydrogen sulfide is poisonous, and when it is mixed with natural gas, causes corrosion in the well. Sweet gas has no detectable hydrogen sulfide, whereas sour gas has detectable amounts. Sweet gas is priced at a premium and sour gas facilities are more expensive to construct and operate to handle the corrosive elements.

Gas-Oil Ratio

The amount of natural gas dissolved in crude oil at the surface is called the producing gas-oil ratio (GOR) and is expressed in cf/bbl. If $Q^o(w, t)$ and $Q^g(w, t)$ represent the oil and gas production associated with well w, then the producing gas-oil ratio for well w is defined as

$$GOR(w,t) = \frac{Q^g(w,t)}{Q^o(w,t)}.$$

If $GOR(w, t) > 5,000$, the well is classified as a "gas" well; $GOR(w, t) < 5,000$ is classified as an "oil" well (Hyne, 1995). Production facility costs are sensitive to the value of GOR and the facilities required to handle large volumes of gas can be significantly more expensive than low/no gas production.

The gross revenues adjusted for the cost of basic gathering, compression, dehydration and sweetening form the base of the royalty:

$$ROY(s,t) = ROY\,(GR(s,t) - ALLOW(s,t)).$$

---

[11] The density, sulfur content, and acidity of a field will usually stabilize after a year or so, but depending on the number of producing zones and degree of commingling, variability over the life of the field can occur.

The total allowance cost is denoted by $ALLOW(s,t)$ and the royalty rate ROY, $0 \leq ROY \leq 1$, depends upon the location and time the tract was leased and the incentive schemes, if any, in effect. The "typical" federal royalty rate in the United States is $ROY = 1/8^{th}$ (12.5%) onshore and $ROY = 1/6^{th}$ (16.67%) offshore. In recent years, the suspension or reduction of royalty payments in certain offshore areas (and subject to specific conditions) has been introduced.

Capital expenditures are the expenditures incurred early in the life of a project, often several years before any revenue is generated, to develop and produce hydrocarbons. Capital expenditures typically consist of geological and geophysical costs, drilling costs, facility costs, construction, transportation, and any other cost required to develop the field. Capital cost may also occur over the life of the project, such as when wells are recompleted into another formation or sidetracked with a horizontal well, artificial lift facilities are installed or facilities are upgraded/replaced, etc. Capital expenditures are decomposed into tangible and intangible costs:

$$CAPEX(s,t) = CAPEX/T(s,t) + CAPEX/I(s,t),$$

where,

$CAPEX/T(s,t) =$ Tangible capital expenditures of structure s in year t,

$CAPEX/I(s,t) =$ Intangible capital expenditures of structure s in year t.

Tangible costs have a useful life greater than one year and a recognizable salvage value, and are depreciated according to federal guidelines. Intangible costs are taken as a tax deduction in the year of expenditure.

Operating expenditures represent the money required for the daily operation of the structure to operate and maintain the facilities; to lift the oil and gas to the surface; and to gather, treat, and transport the hydrocarbons. Many factors influence operating expenditures, including the operator, age and type of field, field equipment, location, efficiency of operating labor, fuel costs[12], wage level, general economic conditions (which affects the cost of oil field services), wear and corrosion. Operating costs and intangible capital costs are typically expensed.

Operating expenditures are frequently described in terms of direct and indirect expense:

$$OPEX(s,t) = OPEX/D(s,t) + OPEX/I(s,t),$$

where,

$OPEX/D(s,t) =$ Direct operating expenditures of structure s in year t,

$OPEX/I(s,t) =$ Indirect operating expenditures of structure s in year t.

---

[12] Oil production is an energy intensive operation, and when fuel prices increase, so does production costs. Gas production is typically more labor intensive with only minor fuel costs.

The general rule for charging costs directly to an operation is that the charges must be for work physically performed at the project site or exclusively for that operation. Costs which are incurred at a distant location for a number of different operations are considered indirect costs or overhead.

Taxable income is determined as the difference between net revenue and operating cost; depreciation, depletion, and amortization; intangible drilling costs; investment credits (if allowed), interest in financing (if allowed), and tax loss carry forward (if applicable). In the United States, state and federal taxes are determined as a percentage of taxable income, usually ranging between 35-50%, and here denoted by the value T, $0 \leq T \leq 1$:

$$TAX(s,t) = T(NR(s,t) - CAPEX / I(s,t) - OPEX(s,t) - DEP(s,t) - CF(s,t) - DECOM(s,t)),$$

where,

$NR(s,t) = GR(s,t) - ROY(s,t) = $ Net revenue of structure s in year t,

$CAPEX / I(s,t) = $ Intangible capital expenditures of structure s in year t,

$DEP(s,t) = $ Depreciation, depletion, and amortization of structure s in year t,

$CF(s,t) = $ Tax loss carry forward of structure s in year t,

$DECOM(s,t) = $ Decommissioning cost of structure s in year t.

The tax and depreciation schedule is normally legislated and will vary across time. In the United States, all or most of the intangible drilling and development cost may be expensed as incurred, whereas equipment cost must be capitalized and depreciated (Gallun et al., 2001). Tax losses in the U.S. may be carried forward for at least three years.

Decommissioning cost represents the expenditures that occur near the end of the life of a structure when the wells are plugged and abandoned; the deck, piles, conductors, and jacket are removed; and the site is cleared of debris. There are many factors that influence the time and cost to decommission an offshore structure, and the primary factors include the physical characteristics and disposition of the structure, the specification of the job, operator preferences, the time of removal, market conditions, and the occurrence and duration of exogenous events.

3.2.4. Economic Indicators:
The purpose of economic evaluation is to assess if the revenues generated by the project cover the capital investment and expenditures and the return on capital is consistent with the risk associated with the project and the strategic objectives of the company (Brealey and Myers, 1991). The present value $PV(s)$ and internal rate of return $IRR(s)$ are the most common measures in the oil and gas industry (Thompson and Wright, 1984) and are computed as

$$PV(s) = \sum_{t=0}^{t_a} \frac{NCF(s,t)}{(1+D)^t},$$

$$IRR(s) = \{D \,|\, PV(s) = 0\},$$

where D is the corporate discount factor and the project is assumed to begin at time $t = 0$ and end at the abandonment time $t = t_a$. The present value provides an evaluation of the project's net worth in absolute terms, while the rate of return is a relative measure used to rank projects for capital budgeting. Economic values are not intended to be interpreted on a stand-alone basis and should be used in conjunction with other system measures and decision parameters.

3.2.5. Typical Cash Flow Patterns: Oil and gas ventures have a great variety of patterns of investment and payout, but most ventures can be decomposed into four basic stages:

> I. Investment/Development
>
> II. Production
>
> III. Marginal Production
>
> IV. Decommissioning

The cash flows in the early years of development are usually large and negative, and so during Stage I, $NCF(s,t) < 0$. Capital expenditures decline quickly after development is complete, and after the field begins to flow, gross revenues turn positive along with $NCF(s,t)$. The cash flow elements peak during the production stage and payout is the point in time when the cumulative net cash flow breaks even. The price of hydrocarbon (and production to some extent) is a stochastic quantity, and so the gross revenues and $NCF(s,t)$ are uncertain throughout the life cycle of the field. The reserves on which the cash flow is derived are finite in nature, and so the elements of the cash flow stream will eventually converge to zero.

Expenses that need to be paid during Stage II and III include royalty, operating cost, and taxes, and typically, capital equipment is fully depreciated before the conclusion of Stage II. The onset of "marginal" production varies with operator and field, and during the later part of Stage II, investment is required to maintain production, but the decision to invest depends on operator preference and economic criteria. Properties in Stage II and III are frequently considered for divestment and joint agreement activity. During Stage IV, the operator will incur a negative outlay when the revenue generating capacity of the structure is exhausted.

3.3. Parameter Forecasting

The physics and engineering of oil and gas exploration, development, and production is the same worldwide, but the geologic and environmental conditions, fiscal regimes, prospectivity, and risk-reward profiles for projects are not homogenous. To construct a model of the time that a structure is expected to be decommissioned, a number of forecasts are required, including

1. Expected reserves, RES (BOE),

2. Projected production forecast of oil and gas, $Q^o(s, t)$ (bbl) and $Q^g(s, t)$ (Mcf),

3. Projected quality of the oil and gas, $g^o(s, t)$ and $g^g(s, t)$,

4. Projected price of oil and gas, $P^o(s, t)$ ($/bbl) and $P^g(s, t)$ ($/Mcf),

5. Projected capital expenditures, $CAPEX(s, t)$ ($) ,

6. Projected tangible and intangible capital expenditures, $CAPEX/T(s, t)$, $CAPEX/I(s, t)$ ($),

7. Projected operating expenditures, $OPEX(s, t)$ ($),

8. Projected tax rate, T (%),

9. Projected decommissioning cost, $DECOM(s, t)$ ($),

10. Estimated discount rate, D (%).

Estimates are based upon the best information available at the time the forecast is made and are considered under the economic conditions projected for the future, including inflation, supply/demand conditions, and technological improvements.

The types of estimates that can be performed depend on the stage of development of the project, the experience of the estimator, and the design and planning information available. Initial cost and production estimates typically fall between "order-of-magnitude" estimates (on the order of 25%-50% accuracy) and "conceptual development plan" estimates (on the order of 15%-25% accuracy) (Mian, 2002). The uncertainty associated with the value of the parameter forecasts will almost always fall within a broad range, and in the worst case, the range itself may be unknown. Estimates made after the field is in production (say, during the mid-point of the life cycle of a field) are usually more reliable, but other sources of uncertainty which are unobservable, such as strategic and technologic factors, enter the analysis at this time.

3.3.1. Reserves: A detailed and realistic field description is the first and most important estimate that is made prior to development. The size, shape, productive zones, fault blocks, drive mechanisms, etc. of the reservoir are estimated by company geologists and engineers as accurately as possible since these factors determine the design capacity of the equipment and structure, the required number and location of wells, and the supporting infrastructure requirements.

Reserves are those quantities of hydrocarbon that are anticipated to be commercially recoverable from known accumulations. Proved reserves are known with reasonable certainty because the field has been defined by appraisal wells. Developed reserves can be produced from existing wells and existing infrastructure, while undeveloped reserves are produced from wells that have not yet been drilled or from existing wells that are "beyond the pipe." Proved reserves are *not* fixed, but rather, depend upon the amount of exploration undertaken, technology and economic conditions, and thus can vary as a result of changes in the external position of these factors (Seba, 2003; Rose, 2001).

Reserves appreciation refers to the *expected* increase in estimates of proven reserves as a consequence of the extension of known pools or discovery of new pools within existing fields. Reserves appreciation, or reserves growth, represents the expected increase in the estimates of original proved reserves of an oil and gas field. Field growth can result from several factors, such as improvements in recovery, physical expansion of the field, better understanding of the reservoir, data re-evaluation, extension drilling, and changes in economic parameters. Changes in reserve estimates for a specific field may be negative as well as positive, but on average, reserve estimates usually grow over time. Field growth is most rapid the first few years after a field is

41

discovered, and later tends to level out at a smaller increment, so that near the mid-cycle of a field, the recoverable reserves are reasonably well-known.

3.3.2. Production Profile:
Many factors impact the rate at which hydrocarbons are produced, but the two primary factors are the geologic conditions and development plan. The geologic conditions at the site – the type and characteristics of rock, depth, thickness, fault mechanisms, hydrocarbon properties – are essentially "fixed," while the development plan – well density, wellbore size, completion techniques, method of production, equipment capacity – represent design parameters. Production rates across fields vary widely because of the variability in these factors.

There is a trade-off in the investment required to produce oil and gas and the production rate achieved. A high production rate requires a large capital investment in the form of the number and type of wells drilled, structure facilities, and the capacity of production equipment. High investment also requires a higher rate of return to justify the increased capital risk, and so the preferences of the operator and their perceived risk-reward tradeoff will determine the design capacity of the field.

Most production profiles can be decomposed into three distinct phases:

Ramp-Up

Production normally builds up over the first few years of production. Following the installation, hookup, and certification of the platform, development drilling is carried out and production started after a few wells are completed. Subsea completions may be used to produce from appraisal wells before full field development.

Plateau

The plateau period represents the maximum rate of production the facilities were designed to handle, pipeline capacity, or contractual constraints. The duration of the plateau is based upon the productivity of the reservoir and the economics of the project.

Decline

After peak production, fields will decline due to the geology and pressure loss at a rate determined by the reservoir drive, investment, and economic conditions. The nature of the decline is characterized through the decline rate.

A reliable production forecast early in the life of a field can only be developed with knowledge of the development plan, reserve estimates and production capacity (Allen and Seba, 2003). Limitations on the availability and accuracy of data constrain the reliability of forecasting. During the mid-point in the life of a field, a different sort of uncertainty arises, since the production profile and the drive mechanisms of the field are now reasonably well understood, but the strategic decisions of the operator are unknown. Will the operator invest additional capital? Will the operator seek a joint operating agreement or divest the structure? Leases are held by a wide variety of working interest owners and are inevitably carved up over time and sold off or subject to a variety of joint venture/farmout type arrangements. Operators purge their portfolios of under-performing and non-core assets on a semi-regular basis, and as properties change hands, the capital expenditures and operating cost structures typically change.

3.3.3. Hydrocarbon Price and Quality: The domestic price of oil and gas is determined by the cost of imports and market conditions. Conversion factors for oil and gas adjust the benchmark price and depend primarily on the API gravity and sulfur content of the produced hydrocarbon. Hydrocarbon prices are a stochastic quantity while production quality is time dependent.

3.3.4. Capital Expenditures: Capital expenditures typically consist of geological and geophysical costs, drilling costs, facility costs, construction, installation, and any other costs required to develop the field (Gallun et al., 2001).

Geological and Geophysical Cost

Geological and geophysical (G&G) costs are pre-drilling exploration costs, and include topological, geological, and geophysical studies. G&G costs may occur before or after the acquisition of working interest in the lease, and for tax purposes, are usually expensed in the year incurred.

Drilling Cost

Drilling time and costs depend on many technical aspects of the well(s) to be drilled, such as the configuration and geometry of the well, type of drilling contract and rig type, well depth and formation complexity. Other factors include the preferences of the operator and performance of the contractor, the weather encountered, and problems associated with the operation. Drilling and development costs are classified as either intangible drilling and development cost or equipment costs.

Development Cost

Three basic options exist to develop reserves: (1) A new platform can be built to drill the development wells, (2) a subsea well can be drilled and tied back to a nearby platform, or (3) an extended reach well can be drilled from an existing platform. The reserve size, distance from existing infrastructure, and time to first production determine which option is the most economical. New platforms and subsea wells are the standard development design, but if a high level of certainty exists that the reserves are in place and within reach of existing infrastructure, then an extended reach well may also be a viable option.

Many different facilities are required to produce oil and natural gas and the exact specification depends, among other things, on the size and distribution of the resource (e.g., broad shallow reservoirs, deep compact reservoirs), fluid variations (e.g., gas-oil ratio, nonassociated gas), meta-ocean conditions (e.g., hurricane risk, wave height, water depth), pipeline infrastructure, local construction infrastructure, and crew accommodation requirements.

Installation Cost

The manufacturing and installation cost of the structure(s) required to develop and produce a field is typically the most significant capital expenditure, ranging between 50-75% of the total costs of the project. Drilling expenditures usually make up the bulk of the remaining cost. In the Gulf of Mexico, total *CAPEX* is frequently assumed to range between $3/BOE-$4/BOE (Johnston, 2000), but these are "zero-type" estimates that are subject to significant uncertainty.

3.3.5. Operating Cost: Direct operating cost can be expressed in terms of subcategories such as production, transportation, maintenance, and other.

Production Cost

Production cost usually contributes the greatest amount to operating cost, but the percentage breakdown varies with the operator, site, and the stage of the project's life cycle. Production costs include the cost to lift and treat (dehydrate and separate) hydrocarbons and to dispose of water, which in turn depends on the capacity of the equipment and the throughput.

Transportation Cost

Transportation costs are related to the transport of oil and/or gas from a field to a refinery or processing facility, an export terminal, or any other point of sale. These costs depend on the throughput, the distance to be covered, and the means of transport. Transportation cost items typically include pump and compressor fuel, tanker rentals (if applicable), pipeline tariffs, and terminal cost.

Maintenance Cost

Maintenance cost is associated with keeping the oilfield equipment and wells in good working condition and production. Maintenance covers material and manpower cost and is usually subdivided into facility and workover categories. Facility maintenance comprises inspection costs, preventative maintenance, and remedial costs. Workover costs occur less frequently and include the costs of well stimulation and repair.

Other Cost

For offshore operations, other direct operating cost items typically include supply boats, helicopters, standby vessels, docking charges, shore base expense, underwater inspections (platforms and pipelines), communications and data transmission, weather services, personnel, small tools and supplies, and equipment standby (e.g., wireline, cementing pumps).

Indirect operating cost items include office expenses, lease supervision, engineering salaries, clerical support, warehouse, management salaries, public affairs, and insurance. Administrative and general overhead may vary significantly among operators, while insurance varies with the cost of replacement and the vulnerability of the insured unit. The method for allocating indirect costs is arbitrary, but prorated and percentage rules are commonly employed (Gallun et al.,

2001). Ernst & Young LLP surveys operators in the U.S. on their average overhead rates per well by producing area and well depth. For offshore wells in the GOM, the monthly median overhead rate per well in 2002-2003 was $35,000 for a drilling well and $3,500 for a producing well. Full cycle operating cost of $2.5/BOE – $3/BOE is frequently assumed for the GOM, and the operating cost in the peak year of production may range from 3-8% total capital expenditures (Johnston, 2000).

3.3.6. Decommissioning Cost: Decommissioning occurs in stages and typically over disjoint time frames. Greatly simplified, following project engineering and cost assessment, federal and state regulatory permits for well plugging and abandonment, pipeline abandonment, and structure removal must be obtained. Approval of the site-clearance methodology (either via trawling/diver/ROV/high-resolution sonar) is given during Structure Removal Permit Application process. Wells are plugged and the facility is prepared for removal, including flushing and cleaning process components, installing padeyes, etc. and then the pipelines are pigged and/or flushed riser-to-riser and riser-to-subsea tie-in, detached from the structure, and capped. Pipelines are normally left in place with the ends buried 3 feet (1 meter) below the mudline. Modules that are to be lifted separate from the deck are removed, the deck is cut and removed, and then the conductors and piles are cut 15 feet (5 meters) below the mudline and pulled. Heavy lift vessels bring the jacket ashore for recycling, sale, or scrap, and in the GOM, the operator may participate in a reefing program. After the structure has been lifted and salvaged, any lost/jettisoned/discarded items are removed from the seabed around the removal site via trawling contractors, ROVs, or divers in order to verify that the area is returned to "pre-lease" conditions (Pulsipher, 1996).

The average total cost to remove 4-pile structures in the GOM has been estimated as $885,000, while the average total cost for an 8-pile structure is $1,344,000. Decommissioning cost can be estimated across each stage of the operation or an aggregate estimate of total cost can be performed based on one or more descriptor variables. For example, if $WD(s)$ represents the water depth at the site (in feet) and $NP(s)$ is a count of the number of piles of the structure, then empirical cost functions such as

$$DECOM(s) = \begin{cases} 2.2WD(s) + 491, & NP(s) \leq 3 \\ 6.2WD(s) + 318, & 3 < NP(s) < 6 \\ 4.2WD(s) + 886, & NP(s) \geq 6, \end{cases}$$

can be employed to estimate decommissioning cost (Kaiser et al., 2003).

3.4. Model Development

Four models are developed to predict structure abandonment time. The models range from a simple, heuristic-based forecast, to a risked, net present value approach. The choice of which of the four models is "best" depends upon the beliefs of the user and the robustness of the model assumptions. Factors not considered in the analysis include the strategic objectives of the operator, the regulatory flexibility associated with decommissioning, and random acts of nature. For a related discussion of decommissioning modeling, see (Kemp and Stephen, 1997).

3.4.1. Model I – Resource Recovery: The simplest "production-based" model of abandonment is derived from an estimate of the time when the expected reserves of the field are depleted. The expected time of abandonment will occur when the forecasted cumulative production equals the reserves expected to be recovered. The resource constraint determines the physical limitation of production since, under the assumptions specified, the reserves will be "depleted" at this time. The expected time of abandonment is designated formally as

$$t_a(\text{I}) = \min\{t' \mid \sum_{t=t_o}^{t'} Q(s,t) \geq RES\},$$

where first production is assumed to start at time t_o. Production is reported on an annual basis, and it is clear that the minimization operator "min" will select the first time when cumulative production exceeds the resource base.

No economic factors influence the result, at least not directly, and both $Q(s,t)$ and RES are estimated quantities. Forecasting $Q(s,t)$ is based on assumptions regarding the decline rate, the time of peak production, and investment decisions. The resource estimate is based on current technology and price levels. After peak production, $Q(s,t)$ is assumed to be a decreasing function of time, and for a given value of RES, the time of first passage will be unique. The uncertainty associated with the analysis depends on the time relative to the production cycle the forecast is performed. If the analysis is performed at the beginning of the life cycle of the field, both $Q(s,t)$ and RES will be significantly more uncertain than if the analysis is performed during the mid-point or near the end of the field's life cycle.

3.4.2. Model II – Threshold Indicators: It is reasonable to assume that "similar" structures will exhibit "similar" conditions[13] near the time of abandonment. If the threshold limit of production and the adjusted gross revenue for structure s is denoted by $\overline{Q}(s)$ and $\overline{GR}(s)$, then the time of abandonment is estimated by

$$t_a(\text{IIa}) = \min\{t \mid Q(s,t) \leq \overline{Q}(s)\},$$

$$t_a(\text{IIb}) = \min\{t \mid GR(s,t) \leq \overline{GR}(s)\}.$$

Hybrid threshold models incorporate the reserves constraint of Model I in the determination of abandonment time; i.e.,

$$t_a(\text{IIc}) = \min\{t' \mid Q(s,t) \leq \overline{Q}(s), \sum_{t=t_o}^{t'} Q(s,t) \leq RES\},$$

$$t_a(\text{IId}) = \min\{t' \mid GR(s,t) \leq \overline{GR}(s), \sum_{t=t_o}^{t'} Q(s,t) \leq RES\}.$$

---

[13] This is explored more completely in Chapter 4.

The inclusion of the reserves constraint ensures that the structure cannot extract more than is available to produce. The reserve constraint is rarely realized in practice, however, since economic and strategic conditions usually dominate removal and divestment decisions.

A structure may reach its economic limit ("first passage") when hydrocarbon prices are in a depressed price-demand state, but if the operator believes stronger prices will prevail in the future, then an abandonment decision is likely to be postponed until the operator can no longer sustain operating losses. To reflect an operator's reluctance to remove a structure at first passage, more stringent conditions can be enforced, such as requiring gross revenues to fall below the threshold two or three (consecutive) years in a row:

$$t_a(\text{IIe}) = \min\{t+1 \mid GR(s,t) \leq \overline{GR}(s), GR(s,t+1) \leq \overline{GR}(s)\},$$

$$t_a(\text{IIf}) = \min\{t+2 \mid GR(s,t) \leq \overline{GR}(s), GR(s,t+1) \leq \overline{GR}(s), GR(s,t+2) \leq \overline{GR}(s)\}.$$

Although production is assumed to decline after peak production, the price of hydrocarbons is a stochastic variable, and thus gross revenues may in fact increase even as production declines.

Model II does not explicitly take into account the capital or operating outlay or the decommissioning cost, but it does include costs in an indirect manner since "similar" structures are being compared, and it is reasonable to assume that similar structures will exhibit similar conditions near abandonment. Model II is considered an improvement over Model I since it incorporates site-specific and historical data into the criteria that determines abandonment time. Model II is also not dependent on the sequence of divestments that occur over the life cycle of the field, since only the abandonment conditions at the end of the life of the structure are considered.

3.4.3. Model III – Negative Cash Flow:
The net cash flow model proxies abandonment time according to the cash flow of the operator. To compute the net cash flow of an investment requires that a detailed life cycle model be developed. Unlike the construction of a cash flow model for investment evaluation, where accurate estimates in the early years of production is most important, abandonment evaluation requires a reliable estimate of cash flow elements to be performed near the end of the production history of the asset where uncertainty is the greatest. Abandonment of a structure is expected to occur when the net cash flow of the asset first turns negative:

$$t_a(\text{IIIa}) = \min\{t \mid NCF(s,t) \leq 0\}.$$

This is frequently referred in the literature as the "economic limit" of the structure.

Similar to the delay decisions incorporated in Model II, an operator may choose to hold an asset for one or more years even if the net cash flow is negative:

$$t_a(\text{IIIb}) = \min\{t+1 \mid NCF(s,t) \leq 0, NCF(s,t+1) \leq 0\},$$

$$t_a(\text{IIIc}) = \min\{t+2 \mid NCF(s,t) \leq 0, NCF(s,t+1) \leq 0, NCF(s,t+2) \leq 0\}.$$

More generally, an operator may abandon a property when a threshold limit on the level of cash flow is reached, say $E > 0$, for $l > 0$ years in a row:

$$t_a(\text{IIId}) = \min\{t + l \mid \bigcap_{t,t+1,\ldots,t+l} NCF(s,t) \leq E\}.$$

Note that Models IIIa, b, c, are a special case of Model IIId for $E = 0$ and $l = 1,2,3$, respectively. E is an operator-defined parameter.

Model III provides a realistic economic assessment from the operator's perspective since it incorporates estimates of the magnitude and timing of future revenue and costs in a standard economic framework. There is a trade-off between modeling the capital and operating expenditures for the asset directly (as in Model III) versus indirectly (as in Model II). In Model II, the structure is compared with historic threshold levels which do not depend on the cash flow position of the operator or the divestment sequence. In Model III, the analyst must either be privy to private information or must make several additional estimates on the magnitude and timing of future costs.

The accuracy and reliability of Model III is plagued by the uncertainty associated with the magnitude and timing of the cost estimates and the uncertainty associated with the strategic objectives of the operator. If an operator divests the structure, which is almost certain to occur at least two or three times during the life of the asset, the cash flow stream of the asset will need to be modified along with updated cost elements for the new operators. Model II and Model III are considered roughly equivalent in their uncertainty, and so a preference for the simpler Model II, prevails.

3.4.4. Model IV – Maximum Net Present Value:
The net present value model refines the net cash flow model by incorporating the time value of money into the relation to establish the probable removal time of the structure. Model IV is based on the cost data employed in Model III, along with two additional variables, the discount rate and decommissioning cost. The discount rate is a user-defined variable, while the decommissioning cost is estimated based on the characteristics of the structure to be removed. In the net present value model, abandonment time is estimated by maximizing the after-tax net present value of the cash flow.

The time to abandon a structure is computed from the optimization model:

$$t_a(\text{IV}) = \{t' \mid \max_{t'=\tau,\tau+1,\ldots} PV(s,[\tau,t'])\},$$

where τ denotes the current time and the time of decommissioning is scheduled for some time t in the future, $t \geq \tau$. The present value relation is computed as

$$PV(s,[\tau,t']) = \sum_{t=\tau}^{t'} \frac{NCF(s,t)}{(1+D)^{t-\tau}}.$$

The operator selects the time of abandonment to maximize the net present value of the investment.

Application of the net present value model is satisfying from an intellectual perspective since the economic criteria is well established and the model incorporates all the essential cash flow elements in the analysis, including the discount factor and cost of decommissioning, but from a practical perspective, the net present value model will almost always result in an unbounded solution. From a present value point-of-view, operators are willing to pay to avoid the cost of decommissioning since the economic benefit of deferred abandonment usually far outweigh the cost of early removal. Other factors play a role in the decision to abandon – such as liability issues, bonding requirements, strategic objectives, and public perception – but these are not factors, per se, in the economic model, and are not incorporated in the analysis.

The cost of decommissioning frequently represents a large negative outlay in the year of removal, and so operators have a strong incentive to delay or transfer this liability. If structure s is removed in the current year, $t = \tau$, the operator foregoes the benefit of remaining production and incurs the cost of decommissioning immediately. The present value of this decision is $PV(s, [\tau, \tau])$. If the operator produces for one additional year, even if production is marginal and does not cover the cost of the operation, and then removes the structure at the end of the year, this yields the present value $PV(s, [\tau, \tau +1])$, which in most cases will dominate $PV(s, [\tau, \tau])$. Since the cost of decommissioning is large and at least one or two orders-of-magnitude larger than the annual operating cost, and since the discount rate can be considered relatively small, say 10-15%, the functional $PV(s, [\tau, t])$ will be an increasing function of time over a rather long time horizon. Thus, the solution to

$$t_a(\text{IV}) = \{t' | \max_{t'=\tau,\tau+1,...} PV(s,[\tau,t'])\},$$

is unbounded, meaning that the time to abandonment will occur far into the future if guided strictly by economic criteria.

The primary weakness of Model IV in predicting the expected age upon abandonment is the unbounded characteristic of the solution and the fact that structural changes in the cost functionals are not considered. The initial development scenario is appropriate for determining the economics of the original investment, but near the mid-life cycle of the field, after the exploration and development cost have been recovered, the operator will consider options for maximizing the value of the field. One of those options is to divest the property or maintain a working interest position in a joint venture arrangement. The potential for divestiture occurs when another company values the asset more than the operator. A field may be considered a candidate for divestment at any time during its life cycle, but the probability of an asset changing hands increases significantly after the mid-life of the field.

3.5. Meta-Modeling Methodology

The impact of changes in system parameters is usually presented as a series of graphs or tables that depict the measure under consideration as a function of one or more variables under a "high" and "low" case scenario. While useful, these approaches are generally piecemeal and the results are anchored to the initial conditions employed. The restrictions associated with geometric and tabular presentations of multidimensional data are also significant; e.g., on a planar graph at most three or four variables can be examined simultaneously. A more general and concise approach to sensitivity analysis is now presented.

The abandonment time of a structure varies with the structural and parametric specification in a complicated manner, but it is possible to understand the interactions of the variables and their relative influence using a constructive modeling approach. The methodology is presented in three steps.

Step 1. For model φ, bound the range of each variable of interest $(X_1, \cdots, X_k) = (d(t),\ P,\ \overline{GR},\ldots)$ within a design interval, $A_i \leq X_i \leq B_i$, where the values of A_i and B_i are user-defined and account for a reasonable range of the uncertainty associated with each parameter. The design space Ω is defined as

$$\Omega = \{\, (X_1,\ldots,X_k) \mid A_i \leq X_i \leq B_i,\ i = 1,\ldots, k \}.$$

Step 2. Sample the component parameters $(X_1, \cdots, X_k)$ over the design space Ω for each model φ, and compute the expected abandonment time, $t_a(\varphi)$, for each parameter selection $(X_1, \cdots, X_k)$.

Step 3. Using the parameter vector $(X_1, \cdots, X_k)$ and computed functional values $t_a(\varphi)$, construct a regression model based on the system data:

$$t_a(\varphi) = f(X_1, \cdots, X_k) = \alpha_0 + \alpha_1 X_1 + \cdots + \alpha_k X_k,$$

where the values of $(\alpha_0, \alpha_1, \cdots, \alpha_k)$ are determined from the simulation analysis.

This procedure is sometimes referred to as a "meta" evaluation since a model of the system is first constructed, and then meta-data is simulated from the model in accord with the design space specifications and system constraints. A good rule of thumb is to sample until the regression coefficients "stabilize." If the regression coefficients do not stabilize, or if the model fits deteriorate with increased sampling, then the variables are probably spurious and linearity suspect. After the regression model is constructed and the coefficient vector $(\alpha_0, \alpha_1, \cdots, \alpha_k)$ determined, if the model fit is reasonable and the coefficients statistically relevant, the value of the $t_a(\varphi)$ functional can be estimated for each model specification for any value of $(X_1, \cdots, X_k)$ within the design space.

3.6. Illustrative Example

Oil and gas ventures have a great variety of patterns of investment and payout, and unless detailed site-specific information is available, cash flow modeling is a generic exercise. The example developed is used to illustrate the manner in which models of the abandonment time can

50

be developed in terms of a meta-evaluation procedure. The field under consideration is labeled XYZ, and the project is dated from the beginning of development expenditures, although money would have also been required for geologic and geophysical cost, leasing cost, and exploration drilling and planning before the decision to proceed with the development was made.

3.6.1. Development Scenario: At the time of development planning, geologists estimated the XYZ field to have between 60-90 million barrels (MMbbl) recoverable reserves spread throughout several geologic zones and total depth ranging between 15,000-18,000 feet. After eight years of production and reservoir modeling, engineers now believe the ultimate recoverable resources to range between 90-110 MMbbl.

The field was developed at a capital cost of $3.5/bbl based on a 100 MMbbl recoverable reserve estimate. The drilling/production facility chosen for development was an 8-pile traditional platform structure designed to handle peak production of 12,000 bbl/day. The gas production of the field is used to supplement on-site power requirements with the remainder reinjected into the field. The hydrocarbons are primarily light, sour oil, with API gravity 42° and 3% sulfur content requiring expensive treatment facilities. There are currently six producing wells and one subsea tieback with production transported to shore through an existing pipeline. There are no other structures on the leasehold.

Based on historic data from similar fields in the area developed with similar technology, the life cycle operating cost are expected to be $3.4/bbl. The capital and operating cost during the first eight years of production are known with certainty, and nearly all the tangible capital cost has now been depreciated. The field is currently in decline, and because the reservoir is isolated, the operator does not anticipate significant additional capital investment. Further, because of the potential liability associated with decommissioning, the operator does not intend to divest the structure.

Reservoir dynamics suggest that a percentage decline equation will govern the future production profile, described by

$$Q(s,t) = (1 - d(s,t))^{t-8} Q(s,8), \ t \geq 9,$$

where $Q(s, t)$ is the annual production rate in thousand barrels per year (Mbbl/year), $Q(s, 8) = 9,180$ Mbbl, and the decline rate $d(t)$ is assumed to vary with time and fall uniformly between 8-13%; i.e., $d(t) \sim U(0.08, 0.13)$.

The price of oil is assumed to be described by a Lognormal distribution with mean $25/bbl and standard deviation $3/bbl: $P \sim LN(25, 3)$ denotes a flat (time-invariant) price profile over the life cycle of the field, while $P(t) \sim LN(25, 3)$ denotes a time-variant price path with prices fluctuating on an annual basis. All capital costs are assumed to be tangible and depreciated according to a straightline 5-year schedule. The investment is assumed to be completely equity financed so interest deductions on the loan need not be considered. No operating taxes besides income tax are considered, and the rate of inflation is assumed constant.

3.6.2. Model I Results: In Model I, the expected abandonment time of structure s is determined from the relation

$$t_a(\mathrm{I}) = \min\{t' \mid \sum_{t=t_o}^{t'} Q(s,t) \geq RES\},$$

where $t_o = 1$ and the ultimate recoverable reserves are assumed Normally distributed with mean 100,000 Mbbl and standard deviation 10,000 Mbbl; i.e., $RES \sim \mathrm{N}(100000, 10000)$. Refer to Table C.1. Abandonment is determined when cumulative production first exceeds the resource estimate. The expected age of the structure upon abandonment is then computed as

$$A(\mathrm{I}) = t_a(\mathrm{I}) - t_o = t_a(\mathrm{I}) - 1.$$

In functional form, the expected age upon abandonment is described by the system parameters through the regression model:

$$A(\mathrm{I}) = \alpha_0 + \alpha_1 \bar{d} + \alpha_2 RES,$$

where $\bar{d}$ is computed as the average value of $d(t)$ over the production profile:

$$\bar{d} = \frac{1}{t_a - 1} \sum_{t=1}^{t_a} d(t).$$

The signs of the regression model are readily hypothesized. A small average decline rate means quick recovery of the reserves; a large decline rate means slow recovery. For a given value of RES, a small decline rate translates to a quick abandonment, while as $\bar{d}$ increases so will the abandonment age. Hence, the coefficient of α_1 is expected to be positive. If the decline rate is held constant, an increase in the reserves RES will extend the field life, and so $\alpha_2 > 0$.

For the design space $\Omega = \{ d(t) \sim \mathrm{U}(0.08, 0.13), RES \sim \mathrm{N}(100000, 10000)\}$ the empirical results of the meta-model yield

$$A(\mathrm{Ia}) = -90.8 + 362.2\bar{d} + 0.00070\,RES, \quad R^2 = 0.62,$$

consistent with expectation. See Table C.2. For any value of $\bar{d}$ and RES within the design space, the regression model can be used to estimate abandonment age; e.g., if $\bar{d} = 10\%$ and $RES = 100,000$ Mbbl, then $A(\mathrm{Ia}) = 15.4$ years. A 1-percentage point increase in the average decline parameter (to $\bar{d} = 11\%$) translates to an abandonment age of 19.0 years, or a 4-year increase in the expected age. Similarly, a 10,000 Mbbl increase in reserves translate to an abandonment age of 22.4 years, or a 7-year increase.

If the shape or size of the design space is changed, the simulation must be recalibrated and the structure equations re-estimated. Adding, deleting, or redefining variables will change the shape of the space, while increasing or decreasing the bounds of the parameters will change the size of the space. For instance, if the model parameter of RES is revised to reflect greater uncertainty on the recoverable reserves, say $\Omega = \{ d(t) \sim U(0.08, 0.13), RES \sim N(100000, 20000) \}$, then Model Ib yields

$$A(\text{Ib}) = -81.1 + 382.3\overline{d} + 0.00066 RES, \quad R^2 = 0.72.$$

In this case, if $\overline{d} = 10\%$ and $RES = 100,000$ Mbbl, then $A(\text{Ib}) = 23.1$ years.

3.6.3. Model II Results: In Model II, the gross revenue is used as a threshold indicator on abandonment, and so the data requirements are expanded to include information on the price of oil and gross revenue. Model II specifies that when the annual gross revenue falls below the threshold limit $\overline{GR}(s)$, the structure will be removed:

$$t_a(\text{II}) = \min\{t \mid GR(s,t) \le \overline{GR}(s)\}.$$

The value of the threshold level $\overline{GR}(s)$ is estimated based on inflation-adjusted historical data near the end of the life of "similar" assets and is assumed to be Uniformly distributed between \$15,000 and \$30,000: $\overline{GR}(s) \sim U(15000, 30000)$.

To compute revenue a price forecast is obviously required. The price of oil can be considered constant over the time horizon of production or a stochastic model of price variation can be employed; e.g., price can be assumed to follow a stochastic price path. If prices are constant over the life cycle of the field, gross revenues will decline in step with production decline; if prices fluctuate, so will the gross revenues.

The functional form of Model II is expressed as

$$A(\text{II}) = \alpha_0 + \alpha_1\overline{d} + \alpha_2 P + \alpha_3\overline{GR},$$

and again the coefficients of the regression model are readily hypothesized. An increase in the average value of $\overline{d}$ will induce a quicker production decline, thereby decreasing the average age of the structure[14], and so $\alpha_1 < 0$. If P increases and all other factors are held constant, gross revenues will increase which will extend the economic limit and delay the time of abandonment, or $\alpha_2 > 0$. As the threshold level $\overline{GR}$ increases, the time of decommissioning will occur sooner, and so $\alpha_3 < 0$.

---

[14] In Model I a fixed reserve base was assumed, implying $\alpha_1 > 0$, while the threshold indicator in Model II implies, $\alpha_1 < 0$.

For the design space $\Omega = \{ d(t) \sim U(0.08, 0.13), P \sim LN(25, 3), \overline{GR} \sim U(15000, 30000) \}$, the results of Model IIa yield

$$A(\text{IIa}) = 53.4 - 229.8\overline{d} + 0.37P - 0.00042\overline{GR}, \ R^2 = 0.97.$$

All the coefficients yield the expected sign as shown in Table C.2. For $\overline{d} = 10\%$, $P = \$22/\text{bbl}$ and $\overline{GR} = \$20,000$, $A(\text{IIa}) = 30.2$ years. A \$1 increase in the price of oil over the life of the field results in an expected age of 30.5 years, or about four additional months of production until abandonment. A \$1,000 increase in the gross revenue threshold limit decreases the expected age to 29.7 years.

The impact of design space modifications is readily explored. For example, if the gross revenue threshold varied on the up-side, indicating that the structure reached its economic limit at a higher level of gross revenue, then we would expect that the time of abandonment would occur sooner since gross revenues are generally a decreasing function of time after peak production. For Model IIb,

$$\Omega = \{ d(t) \sim U(0.08, 0.13), P \sim LN(25, 3), \overline{GR} \sim U(15000, 50000) \},$$

the simulation yields

$$A(\text{IIb}) = 50.8 - 208.9\overline{d} + 0.36P - 0.00034\overline{GR}, \ R^2 = 0.96;$$

e.g., for $\overline{d} = 10\%$, $P = \$22/\text{bbl}$ and $\overline{GR} = \$40,000$, $A(\text{IIb}) = 24.2$ years.

The impact of structural modifications on the model can also be explored. For instance, if the hydrocarbon price is assumed to vary annually as $P(t) \sim LN(25, 3)$, and the average price over the time horizon is denoted $\overline{P}$, then the empirical results indicate that the Model IIc coefficients remain fairly stable but lose some of their statistical significance – refer to the fixed term and the price coefficient in Table C.2 – and compare the results with Model IIb:

$$A(\text{IIc}) = 51.5 - 206.9\overline{d} + 0.29\overline{P} - 0.00034\overline{GR}, \ R^2 = 0.95;$$

e.g., for $\overline{d} = 10\%$, $\overline{P} = \$22/\text{bbl}$ and $\overline{GR} = \$40,000$, $A(\text{IIc}) = 23.6$ years.

3.6.4. Model III Results: To estimate the net cash flow position of the operator, a host of new variables are required. In addition to the parameters $d(t)$ and $P(t)$, we now also require the royalty and tax rate, operating expenditures, capital expenditures, tangible and intangible cost decomposition, and the depreciation schedule. As the problem was formulated, the value of these parameters is known for the first eight years of production, but during the decline stage additional structural assumptions will be required.

The annual operating expenditures for the 8-pile, 6-well structure are assumed to be given by the relation

$$OPEX(s,t) = \$10{,}020 + \$1.6Q(s,t),$$

for $Q(s,t)$ described in Mbbl/year. This relation is based on historic data of the field and an assessment of similar structures in the region. To reflect changes that may occur in the value of the operating costs, a perturbation factor $k \sim U(0.9, 1.3)$ is applied to the annual value of $OPEX(s,t)$. For $k < 1$, operating cost would be smaller than the historic relation, while for $k > 1$, the operating costs would exceed historic rates. The value of k is assumed constant throughout the cash flow cycle, but it is easy to allow k to vary annually, in which case the average measure $\bar{k}$ would be the output variable of interest. The royalty and tax rate are assumed to be Uniformly distributed with $ROY \sim U(0.10, 0.20)$ and $T \sim U(0.10, 0.20)$, and the threshold value of the net cash flow cut-off is $E \sim U(4000, 8000)$.

The functional form of Model III is expressed as

$$A(\text{III}) = \alpha_0 + \alpha_1 \bar{d} + \alpha_2 \bar{P} + \alpha_3 ROY + \alpha_4 k + \alpha_5 T + \alpha_6 E.$$

The expected signs of the coefficients $\alpha_1 < 0$ and $\alpha_2 > 0$ follow from the discussion for Model II. As the royalty and tax rate increase, the net cash flow position of the operator will be negatively impacted, and so we expect $\alpha_3 < 0$ and $\alpha_5 < 0$. The coefficient α_4 reflects the influence of perturbations to the operating expenditures, so that as k increases, operating expenditures increase, again negatively impacting the net cash flow position of the operator. Similarly, as the value of the net cash flow threshold E is raised, structures will be abandoned earlier, and we expect $\alpha_4 < 0$ and $\alpha_6 < 0$.

The net cash flow projection for the field is computed according to the framework previously described. The gross revenues are determined as the product of the production and price trajectory, and the net revenue is determined after the royalty rate ROY is specified. The values for $CAPEX$ and $OPEX$ and the depreciation schedule are known for the first eight years of the field's life and are extrapolated thereafter. The tax is determined after the tax rate T is specified.

Four different scenarios are considered using the parameter values shown in Table C.3. The design space common to each model is given as

$$\Omega = \{\, d(t) \sim U(0.08, 0.13),\ P \sim LN(25, 3),\ ROY \sim U(0.10, 0.20),$$
$$k \sim U(0.9, 0.13),\ T \sim U(0.30, 0.50),\ E \sim U(4000, 8000)\}.$$

The results of the regression models are depicted in Table C.4. For Model IIIa,

$$A(\text{IIIa}) = 59.2 - 206.1\bar{d} + 0.17\bar{P} - 12.6ROY - 4.8k - 7.3T - 0.00071E,\ R^2 = 0.95.$$

This model is based on a sampling scheme involving 1,000 iterations. For $\bar{d} = 10\%$, $\bar{P} = \$25/\text{bbl}$, $ROY = 16.67\%$, $k = 1.1$, $T = 40\%$, and $E = \$8{,}000$, the meta-model yields an expected abandonment age $A(\text{IIIa}) = 26.9$ years. For 5,000 model iterations in Model IIb, the regression coefficients remain fairly stable and generally increase in significance, with the difference in the numerical result between the two models imperceptible: $A(\text{IIIb}) = 26.4$ years.

Alternative decision criteria can be adopted within the analytic framework. Delay can be incorporated in the model by adopting the decision rule

$$t_a(\text{IIIc,d}) = \min\{t + l \mid \bigcap_{t, t+1, t+2} NCF(s,t) \le E\}.$$

In Model IIIc and Model IIId, the net cash flow elements must fall below E for two and three consecutive years before the operator decides to abandon. Obviously, additional constraints on the production profile will increase the expected age of the structure, and so the relevant question concerns the relative impact of the constraint. If the production decline dominates the hydrocarbon price volatility near the time of abandonment, then the incremental impact on the average age is expected to be about one year or so per additional constraint. On the other hand, if the volatility of the hydrocarbon price is a dominant factor, then we would expect the impact to deviate from the one year increment. For $\bar{d} = 10\%$, $\bar{P} = \$25/\text{bbl}$, $ROY = 16.67\%$, $k = 1.1$, $T = 40\%$, and $E = \$8,000$, Model IIIc, d yield

$$A(\text{IIIc}) = 27.1 \text{ years}, \quad A(\text{IIId}) = 28.2 \text{ years},$$

suggesting that production decline is the dominating factor.

3.7. Limitations of the Analysis

Significant sources of uncertainty underlie all models of decommissioning, and the framework described herein only hints at the complexity involved. Additional sources of uncertainty are now described.

3.7.1. Private Uncertainty: The primary sources of private uncertainty include geologic uncertainty, production uncertainty, investment uncertainty, and strategic uncertainty. Some forms of uncertainty are observable and quantifiable (e.g., price), while other forms are quantifiable but unobservable due to their proprietary nature (e.g., geologic). The most difficult forms of uncertainty to model are strategic decisions that are neither observable nor readily quantifiable.

3.7.2. Scale Economies: Operators who divest or farm out a structure induce a structural change in the operating cost of the asset. If O_1 represents the seller and O_2 the buyer, then the typical structural change would be

$$OPEX(s, t, O_2) \le OPEX(s, t, O_1),$$

where $OPEX(s, t, O_i)$ represents the operating cost of structure s in year t for operator O_i. Buyers reduce the cost to operate a structure through scale economies and by maintaining low overhead.

By bundling structures in a group $\{s_1, \ldots, s_k\}$ and servicing the needs of the group as a unit, scale economies can frequently be achieved, such that

$$OPEX(s_1, \ldots, s_k; t, O) \leq \sum_{i=1}^{k} \sum_{j=1}^{l} OPEX(s_i, t, O_j),$$

providing the asset a new lease on life. The decision to invest in structure s_{k+1} when a bundled unit $\{s_1, \ldots, s_k\}$ already exists is an economic decision determined by the incremental benefits of adding the production of s_{k+1} versus the incremental costs of operation and decommissioning. If structure s_{k+1} is in the same geographic area as other properties then the scale economies may provide residual benefit to the owner.

Similar to the structural changes that occur under divestment, operators can reduce the overall cost to decommission structures on a lease through timing and scale economies. Again, by bundling structures in a group $\{s_1, \ldots, s_k\}$ and performing the removal at one time, economies are frequently achieved through more favorable contract terms, reduced mobilization/ demobilization fees, etc. so that:

$$DECOM(s_1, \ldots, s_k, t) \leq \sum_{i=1}^{k} \sum_{j=1}^{l} DECOM(s_i, t_j).$$

Niche operators can act faster and are more operationally flexible than large independents or majors, and this flexibility has value that is expressed in various ways; e.g., niche players can wait until the market rates for construction barges are competitive to perform deconstruction activities.

3.7.3. Regulatory Uncertainty: Typically, a lease is terminated when production on the lease ceases, but if the operator intends to re-work wells or is pursuing drilling activity on the lease, or the lease contains an active pipeline, conditions may warrant the MMS to grant an extension of the lease termination. Since several structures may be contained on a lease, it is only when production from the *last* productive structure on the lease ceases that *all* the structures are required to be removed.

3.7.4. Random Events: Random acts of nature (e.g., see (Daniels, 1994)) also influence the ability to predict removal times, but because the frequency of such events is relatively small, these occurrences do not play a large role in aggregate removal patterns.

3.8. Conclusions

Four models were developed to model the decommissioning time of an offshore oil and gas structure, and the limitations, refinements, and extensions of each model were discussed. The models were then implemented on a generic field development plan to illustrate the simulation methodology and the manner in which the system variables interact. A meta-modeling methodology was used to construct functionals that describe how the age of the structure upon abandonment is related to various system parameters.

The high degree of uncertainty and the large number of factors associated with structure removal suggest that simple models can capture the essence of a removal forecast in a manner that is comparable to more sophisticated methodologies. Academics would probably favor the more sophisticated constructions and consider their application to be more "reliable" predictors of events, but this is not necessarily the case. Under conditions of unknown uncertainty, the results of simple models should not be discounted out-of-hand in favor of more elegant (but just as uncertain) "economic" models. When choosing between methodologies that are constrained by high levels of uncertainty and factors that cannot be adequately modeled, simple models are the preferred approach.

CHAPTER 4: A STATISTICAL ANALYSIS OF THE ECONOMIC LIMIT OF OFFSHORE HYDROCARBON PRODUCTION

4.1. Introduction

The economic life of a structure is defined as the time at which the production cost of the structure is equal to the production revenue. Economic life is normally difficult to determine directly, since full and accurate economic data are often not available on an individual structure basis, and factors such as hydrocarbon price, operational expenditures, investment decisions, and strategic objectives contribute to the uncertainty. Toward the end of the lifetime of most structures, the capital expenditures and depreciation are generally negligible and the operating cost is the primary expense element. When the gross revenue falls below the operating cost, the operator will usually shut down production and consider available divestment or decommissioning options.

The economic limit of structure s, $t_e = t_e(s)$, is defined as the time when the gross revenue of the structure, $GR(s, t)$, equals production cost, $C(s, t)$:

$$t_e = \{t \mid GR(s, t) = C(s, t)\}.$$

Gross revenue $GR(s, t) = P(t)Q(s, t)$ is an observable quantity, while production cost, $C(s, t)$, is generally unobservable. Operators on federal leases are required to report production data to the government on a well basis, and so the gross revenue and royalty stream of a structure can be estimated, but the revenue stream is only half of the equation and the cost data still needs to be inferred. As most casual observers of the oil/gas industry are aware, it is rare indeed when the net cash flow from a real asset is available outside the firm, and rarer still if it is made public. Cash flow and cost information is proprietary. Fortunately, a proxy of the economic limit of a structure can be inferred by observing the production and gross revenue during the structure's last year of production.

The last year of production of structure s, $t_{lp} = t_{lp}(s)$,

$$t_{lp} = \min\{t \mid Q(s, t+1) = 0\},$$

represents a "snapshot" of the economic conditions specific to the field, structure, operator, technology and time of operation. The production and gross revenue at t_{lp}, $Q(s, t_{lp})$ and $GR(s, t_{lp})$, serves as a proxy of the economic limit and can be used to infer the threshold operating cost of the structure. The link between the gross revenue and operating cost threshold is not perfect, but if properly normalized and interpreted, is believed to be a reasonably good indicator of the economic limit of a structure.

The purpose of this chapter is to provide a statistical assessment of the economic limit of offshore structures in the GOM. The model development is presented, followed by statistical relations of the production and revenue threshold levels. Conclusions complete the chapter.

4.2. Model Development

The economic limit of a structure is estimated following four steps:

Step 1. Estimate structure production;

Step 2. Define lease categorization;

Step 3. Compute production and gross revenue thresholds; and

Step 4. Define factor variables.

A description of each stage is now provided.

4.2.1. Production Allocation: Each structure is identified with a lease, and in theory, each well is associated with a (unique) structure. In practice, however, not every well in the MMS database is linked to a structure. Wells are grouped into two disjoint categories – assigned $\{w^a\}$ and unassigned $\{w^u\}$ sets. A well that is a member of $\{w^a\}$ maintains a structure assignment while wells in $\{w^u\}$ require assignment. Unassigned well production is allocated to the nearest lease structure. For $w^u \in l$, assign w^u to structure $s \in l$ based upon the criteria,

$$w^u \leftrightarrow \{s \mid \min_{s \in l} d(w^u, s)\},$$

where $d(w^u, s)$ represents the distance from well w^u to structure s and the minimization is performed with respect to all structures contained on the same leasehold l as the well w^u. The annual production of structure s is modified as follows:

$$Q(s^a, t) = Q(s, t) + Q(w^u, t).$$

4.2.2. Lease Categorization: Structure data is delineated along two dimensions – the number of structures removed and the number of producing structures on a lease at the time of removal. It is necessary to disaggregate structures in this manner for two reasons:

1. Federal regulations allow structures on a lease to remain idle (nonproductive) as long as the lease is productive. Producing leases can thus "hold" structures idle for a number of years.

2. Structures that exist in close proximity to other structures (on the same lease or adjacent leaseholds) are more likely to exhibit economies of scale in operation and shared expenses, and at the time of removal, reduced decommissioning cost.

Four categories are employed to disaggregate structure/lease data:

I. One structure removal, no other producing structure on lease at time of removal;

II. One structure removal, at least one producing structure on lease at time of removal;

60

III. Two or more structure removals, no other producing structure on lease at time of removal;

IV. Two or more structure removals, at least one producing structure on lease at time of removal.

Every structure removed in the GOM is an element of one and only one lease category, and since the four categories are disjoint, it is clear that the aggregation strategy represents the universe of all removals. Category I represent structures with one and only one structure on the leasehold at the time of removal. If one structure is removed while one or more structures are active on the lease, the structure is classified in category II. Two or more structures may also be removed at the "same" time, or nearly the same time, typically within a month or so of one another. Structures on leases with two or more structures removed but no other producing structures on the lease are classified in category III. Category IV denotes two or more removals on a leasehold with at least one producing structure at the time of removal.

4.2.3. Threshold Limits: The annual production and gross revenue streams of structure s are computed at time t_{lp} and 1, 2, …, k years before last production, as

$$\{Q(s, t_{lp}), Q(s, t_{lp}-1), …, Q(s, t_{lp}-k)\},$$

$$\{GR(s, t_{lp}), GR(s, t_{lp}-1), …, GR(s, t_{lp}-k)\}.$$

The production and gross revenue streams at time t_{lp} are referred to as a "threshold" limit, and denoted by $\overline{Q}(t_{lp}) = Q(s, t_{lp})$ and $\overline{R}(t_{lp}) = GR(s, t_{lp})$. Production and gross revenue levels at/near the year of last production represents a threshold for economic operations under conditions specific to the field, structure, operator, technology, and time of operation. The production and gross revenue levels represent a "snapshot" of the structures state in the year of last production, and presumably, conditions that approximately describe the economic limit.

Threshold indicators $\overline{Q}(\Gamma,t)$ and $\overline{R}(\Gamma,t)$ for category Γ at time t can be computed in several ways. An averaging process is the most common:

$$\overline{Q}(\Gamma,t) = \frac{1}{\#\Gamma(t)} \sum_{s\in\Gamma} Q(s,t),$$

$$\overline{R}(\Gamma,t) = \frac{1}{\#\Gamma(t)} \sum_{s\in\Gamma} GR(s,t),$$

where $\#\Gamma(t)$ denotes the number of elements (structures) in category Γ at time t.

4.2.4. Factor Description: Denote $t_i = t_i(s)$ as the time structure s is installed and $t_r = t_r(s)$ as the time the structure is removed. The idle time (idle age) of structure s is the amount of time that has elapsed between the time the structure was removed and the last year of production,

$$IDLE = t_r - t_{lp}.$$

The profile $Q(s, t)$ is used to delineate three phases of production. If peak production,

$$Q^* = Q^*(s,t) = \max Q(s,t),$$

occurs in year t_p,

$$t_p = \min\{t \mid Q(s,t) = Q^*\},$$

then the peak production period is defined as the time interval when production exceeds αQ^*, where the value of α, $0 < \alpha < 1$, is user-defined and selected near the upper bound of the interval; e.g., $\alpha = 0.8$. The plateau production period is defined by the time interval $[t_c, t_d)$, where

$$t_c = \min\{t \mid Q(s,t) \geq \alpha Q^*, 0 < \alpha < 1\},$$
$$t_d = \max\{t \mid Q(s,t) \geq \alpha Q^*, 0 < \alpha < 1\}.$$

The ramp-up period is defined by $[t_i, t_c]$. The decline period is defined by $[t_d, t_{lp}]$.

The decline rate $d(s, t)$ of structure s in year t, $t_d \leq t \leq t_{lp}$, is defined as

$$d(s,t) = \frac{Q(s,t-1) - Q(s,t)}{Q(s,t)}.$$

Usually, $d(s, t) \geq 0$ but investment decisions, maintenance, production problems, weather and other events may change the sign of the decline rate for one or more years. The average decline rate of structure s over $[t_d, t_{lp}]$ is computed as

$$DEC(s) = \frac{1}{t_{lp} - t_d} \sum_{t=t_d}^{t_{lp}} d(s,t).$$

At the time of development, the peak production to expected reserves ratio, $Q^*/E[RES]$, serves as a proxy for the maximum efficient production rate. To obtain a high $Q^*/E[RES]$ ratio, the operator will need to have a large number of producing wells and adequate production equipment to handle the volumes of oil and gas produced. A low $Q^*/E[RES]$ ratio provides an indirect indication that an operator has chosen to drill less wells and produce longer. Fewer wells require smaller production, processing, and transportation facilities; less operating personnel; reduced financing cost, and presumably, lower operating expenditures. At the time of last production, the expected value of the recoverable reserves, $E[RES]$, is a deterministic and known quantity, RES, computed as

$$RES(s) = \sum_{t=t_i}^{t_{lp}} Q(s,t).$$

4.3. Descriptive Statistics

4.3.1. Data Source: The data for this analysis was obtained through the MMS Technical Information Management System database. The sample set contains over 2,000 structures removed in the GOM over the past two decades, and after eliminating structures that have never produced and structures with missing and/or ambiguous data, the final data set included 1,790 elements. Only structures that have been removed from operation are under consideration. It is clear that production and revenue thresholds can be calculated with a reasonably high level of accuracy since production and price forecasting is not required. Uncertainty arises in the calculation of $Q(s, t)$ and $GR(s, t)$ because of the nature of well assignments, and since the quality of production is not considered, but the exclusion of these factors are not expected to have a significant impact on the results.

4.3.2. Lease Categorization Statistics: Summary statistics for the average production and revenue threshold levels ($\overline{Q}(t_{lp})$, $\overline{R}(t_{lp})$), idle time (*IDLE*), design ratio (*Q*/RES*), and total production (*RES*) are presented in Table D.1 according to lease categorization and structure type. The number of elements within each category is denoted by n.

Structures on leases with no other infrastructure at the time of removal cannot be held by lease production because federal regulations generally require removal within one year after lease production terminates. The average production/revenue thresholds for categorization I structures should therefore be the highest among all lease categories because scale economies cannot be used to minimize operational cost. The average production and revenue threshold values for lease category I, $\overline{Q}(t_{lp}) = \$57,000$ and $\overline{R}(t_{lp}) = 734,000$ BOE, exceed category II, III, and IV levels as expected, and this dominance is maintained across each structure type. The average idle time of categorization I structures should be the smallest among the four lease categories, and indeed this is also the case. The average idle time of structures removed on leases with no infrastructure at the time of removal is *IDLE* = 2.3 years which is nearly three times less than the average idle age of categorization II structures.

For leases with more than one producing structure at the time of removal, we would expect production/revenue thresholds to decrease because of the potential for scale economies, while average idle times should increase because structures can be held by lease production, and this is supported by the empirical data. Leases with two or more structures removed with no other producing structures at the time of removal are more "similar" to category I structures, and so the difference between the production and revenue thresholds and idle time is expected to be smaller. For leases with at least one producing structure at the time of removal, economic theory suggests that

$$\overline{Q}\,(\text{II}) \geq \overline{Q}\,(\text{IV}),$$

$$\overline{R}\,(\text{II}) \geq \overline{R}\,(\text{IV}),$$

$$IDLE(\text{II}) \leq IDLE(\text{IV}).$$

In this case, the production/revenue thresholds relations are not supported by the empirical data, but the idle time behaves as expected.

63

Summary statistics for Q^*/RES and RES provide a quick description of the field characteristics. The value of Q^*/RES range broadly between 20-60%, with gas fields produced with Q^*/RES as high as 40-50%. The average decline rate will often equal or exceed the Q^*/RES ratio; i.e., $DEC \geq Q^*/RES$, and this is generally supported by the empirical data. Q^*/RES is a decreasing function of structure complexity for each lease categorization, while the recoverable reserves is an increasing function of structure complexity. In symbolic form,

$$Q^*/RES(C) > Q^*/RES(WP) > Q^*/RES(FP),$$

$$RES(C) \leq RES(WP) \leq RES(FP).$$

Structure complexity serves as a proxy of the development plan, expected size and decline characteristics of the field, and this is reflected by the relational forms.

4.3.3. Average Production and Revenue Threshold Levels:
Categorization of the data according to lease characteristics, structure type, and water depth is presented in Tables D.2-D.9. The number of elements within each category in the year of last production is shown in the third column of each table. Because some fields have a very short lifetime, the number of elements within each category will decrease with the time from last production (so as one moves horizontally across each row, the size of the sample set will decrease).

The value of the average production thresholds is fairly uniform across water depth and structure type in Table D.2. Well protectors in the 101-200 feet water depth category appear to be an exception due to the presence of a small number of large fields. The average production threshold ranges between 43,000 - 57,000 BOE, meaning that on average, when the annual production from an offshore structure falls within this range the structure is very near its economic limit. The gross revenue thresholds shown in Table D.3 exhibit greater variability than the Table D.2 elements since the production data is modified by (inflation-adjusted) hydrocarbon prices. Excluding well protectors in the 101-200 feet water depth category, the average revenue threshold ranges between $556,000 - $797,000.

In Tables D.2-D.9, the general trends of the production and revenue thresholds are reasonably consistent within each categorization, namely, $\overline{Q}(t_{lp} - t)$ and $\overline{R}(t_{lp} - t)$ are increasing functions of time and the production/revenue thresholds of fixed platforms generally dominate the production of simpler, smaller structures, such as caissons and well protectors,

$$\overline{Q}(C, t_{lp} - t) < \overline{Q}(WP, t_{lp} - t) < \overline{Q}(FP, t_{lp} - t),$$

$$\overline{R}(C, t_{lp} - t) < \overline{R}(WP, t_{lp} - t) < \overline{R}(FP, t_{lp} - t),$$

for $t \geq 1$. The dominance of well protectors and fixed platforms is apparent after just a few years, while the distinction between well protectors and fixed platforms is not always apparent.

As the number of elements within each category decrease, inconsistencies arise in lease categories III and IV, but this is to be expected considering reductions in the sample size and diversity of the fields under consideration. Some indication that the production/revenue threshold is structure invariant is provided, however, since $\overline{Q}(C, t_{lp}) \approx \overline{Q}(WP, t_{lp}) \approx \overline{Q}(FP, t_{lp})$ and $\overline{R}(C, t_{lp}) \approx \overline{R}(WP, t_{lp}) \approx \overline{R}(FP, t_{lp})$, at least approximately, across lease categorizations. There is also a weak trend for the production and revenue thresholds to increase with water depth:

$$\overline{Q}(ST, 0-100') < \overline{Q}(ST, 101-200'),$$
$$\overline{R}(ST, 0-100') < \overline{R}(ST, 101-200') < \overline{R}(ST, 201-400').$$

4.4. Conclusions

The economic limit of an offshore structure is important from an operational perspective and provides insight into the nature of decommissioning activities. No modeling framework is perfect, however, and the best a model can do is to provide insight and ensure an interpretation supported by empirical data. The unique nature of the economic limit of structures drives the observed variability in the data set, and since the category descriptors are constrained and finite, the impact of unobservable factors on the model results may be significant. Many factors impact the economic limit of a structure and it is not possible to enumerate all the factors in modeling, but statistical analysis allows insight to be developed.

CHAPTER 5: FORECASTING THE REMOVAL OF OFFSHORE STRUCTURES IN THE OUTER CONTINENTAL SHELF

5.1. Introduction

Offshore structures combine capital, labor, materials and fuel to produce hydrocarbons, and operate under the physical laws and engineering specification of the system, economic principles which determine the design and commerciality of production, and man-made rules governing operation and decommissioning activities. Significant interrelationships exist between the physical laws by which a system operates and the commercial rules and regulations established for the system.

Structures are installed to produce and process hydrocarbons, and when the time arrives that the cost to operate a structure – or more commonly, a group of structures – exceeds the income from the hydrocarbons under production, the structure(s) exists as a liability instead of an asset and becomes a candidate for divestiture or decommissioning. Over 2,200 structures have been removed from the GOM over the last half-century, and over the past decade, 125 structures on average have been removed annually (Table E.1).

At the end of 2003, there were 2,175 active (producing) structures, 898 idle (non-producing) structures, and 440 auxiliary (never-producing) structures on 1,356 active leases; and 329 idle structures and 65 auxiliary structures on 273 inactive leases (Table E.2). An active structure produced hydrocarbons in the year 2003, while an idle structure once produced hydrocarbons, but has not been productive over the past year. An auxiliary structure has never produced hydrocarbons, but serves in a support role, as a quarter facility, flare tower, storage platform, etc. A total of 2,175 active structures, 1,227 idle structures, and 505 auxiliary structures, or 3,907 total structures, currently reside in the GOM. More than one-half of producing structures reside on leases with one or more active structure.

The distribution of structures according to structure type, water depth, and planning area is shown in Table E.3. A few dozen deepwater structures classified as compliant tower, spar, tension-less platform, mini tension-leg platform, and semisubmersible are not considered in Table E.3 because they are small in number, relatively recent installations, and are expected to serve as deepwater hubs for hydrocarbon production for many years to come.

Federal regulations specify the manner in which structures are removed from the GOM and idle iron is maintained. Changes in the regulation, or re-interpretation of removal requirements, will change the nature of decommissioning services, changing the number of inputs (i.e., amount of service equipment, timing of services) used in decommissioning, which will impact the overall cost and development options available to the operator.

The purpose of this chapter is to develop a general framework to model the removal process that governs structure abandonment and decommissioning in the Gulf of Mexico. The outline of the chapter is as follows. The general modeling framework is first formalized, followed by model parameterization and the results of the analysis. A brief discussion of the limitations of the analysis is also presented. Conclusions complete the chapter.

5.2. Model Framework

5.2.1. General Methodology:
The methodology adopted in this paper follows a five-step process. For structure s and time t,

Step 1. Forecast production profile, $Q(s, t)$,
Step 2. Forecast revenue profile, $R(s, t)$,
Step 3. Estimate abandonment time, $t_a(s)$,
Step 4. Estimate removal time, $t_r(s)$, and
Step 5. Estimate removal cost, $C(s)$.

5.2.2. Production Model:
Hydrocarbon production profiles are generated using the iterative functional,

$$Q(s, t + \alpha(s)) = (Q(s, t + \alpha(s) - 1) + \beta(s)) [1 - d(s, t + \alpha(s))],$$

where $d(s, t)$ represents the decline parameter selected from a Normal distribution with mean, $DEC(s)$, and standard deviation, σ_{DEC}. For structures that have already reached peak production, $\alpha(s) = \beta(s) = 0$; for structures that have not yet achieved peak production, $\alpha(s)$ and $\beta(s)$ are used to adjust the profile to the expected peak production time and rate. The values of $DEC(s)$, σ_{DEC}, $\alpha(s)$, and $\beta(s)$ are derived from the statistical analysis of structures previously removed from the Gulf of Mexico.

5.2.3. Revenue Model:
The gross revenue stream is determined from the relation,

$$R(s, t) = g(s)P(t)Q(s, t),$$

where the conversion factor, $g(s)$, depends on the gravity and sulfur content of oil, and the amount of impurities, condensate, and hydrogen sulfide of natural gas. The hydrocarbon price, $P(t)$, is based on a reference benchmark determined from historical data.

5.2.4. Abandonment Time:
A structure will cease production when its economic limit is reached, that is, when the revenue stream converges to its threshold level:

$$t_{a,Q}(s) = \min\{t \mid Q(s,t) = \varepsilon_Q(s)\},$$

$$t_{a,R}(s) = \min\{t \mid R(s,t) = \varepsilon_R(s)\}.$$

The abandonment time, $t_{a,Q}(s)$, is associated with the production threshold level, $\varepsilon_Q(s)$; the abandonment time, $t_{a,R}(s)$, is associated with the revenue threshold level, $\varepsilon_R(s)$. The threshold levels, $\varepsilon_Q(s)$ and $\varepsilon_R(s)$, are empirically derived and vary with characteristics such as water depth, structure type, and hydrocarbon production, and the number of structures on the lease. It is possible to construct general expressions of the threshold limit as a function of these and other factors.

5.2.5. Removal Time: To estimate removal time, a rule-based policy is assumed to govern and approximate operator behavior. If structures $\{s_1, \cdots, s_k\}$ exist on lease l and are held until production from the *last* structure ceases, then the time in which *all* the structures on the lease are removed is determined from the relation:

$$t_r(s_i) = \max_{i=1,\ldots,k}\{t_a(s_i)\} + 1, \, i = 1,\ldots, k;$$

e.g., if one structure exists on the lease, $s \in l$, then

$$t_r(s) = t_a(s) + 1,$$

while if two structures exist on the lease, $\{s_1, s_2\} \in l$, then

$$t_r(s_1) = t_r(s_2) = \max\{t_a(s_1), t_a(s_2)\} + 1,$$

and so on for three or more structures.

A lease can "hold" structures idle without violating federal guidelines as long as the lease remains producing. Operators may remove idle structures on producing leases early, and indeed, depending upon decommissioning schedules or other preferences, it may be economic for operators to remove select idle structures, but these provisions are not considered. By excluding the possibility that idle structures will be removed earlier than required by law, we are modeling a latest possible removal scenario.

5.2.6. Removal Cost: The cost to remove structure s at time $t_r(s)$ is determined as the present value of the removal cost,

$$C(s) = \frac{C(s)}{(1+d)^{t_r(s)-\tau}},$$

where $C(s)$ denotes the decommissioning cost; d is the discount rate, $0 < d < 1$; τ is the observation time; and $t_r(s)$ is the removal time. The decommissioning cost is described through empirically-derived functions and is assumed constant over time. Potential cost savings associated with scale economies are not considered, and since decommissioning is generally a low-technology, low-margin operation with no significant barriers to entry, technological progress is not expected to have a significant impact on future cost drivers.

5.3. Model Parameterization

Initialization

The GOM currently contains over 1,356 active leases and 273 inactive leases. The number of active, idle and auxiliary structures on active leases was depicted in Table E.2; the number of idle and auxiliary structures on inactive leases is shown in Table E.4. Structures on inactive leases form an inventory of platforms that are expected to be removed in the near future, unless they serve a useful economic purpose, or other special circumstance apply. The 245 structures on inactive leases are not considered. The 440 auxiliary structures on active leases are assumed to be removed when production on the lease on which they reside ceases. The remaining 2,175 active structures and 898 idle structures on active leases are removed according to the latest possible removal scenario.

Production Model

For each producing structure in the GOM, the decline parameter, $d(s, t)$, is sampled each year from the Normal distribution, $N(DEC(s), \sigma_{DEC})$, to forecast the production profile. The value of $DEC(s)$ is computed from historical data of structures removed in the GOM; e.g.,

$$DEC(s) = -14.8 - 0.035ST - 0.019GORV - 0.0104NW + 0.0017O/G + 0.00771VINT,$$

where ST = structure type (ST = 0, caisson; ST = 1, otherwise), $GORV$ = gas-oil ratio variation, NW = number of wells, O/G = oil/gas structure (O/G = 0, oil; O/G = 1, gas), and $VINT$ = vintage (year structure installed). The values of σ_{DEC}, $\alpha(s)$, and $\beta(s)$ are similarly derived.

Revenue Model

The quality and sulfur content of the hydrocarbon stream is field and time dependent, and for convenience, is assumed uniform across GOM fields. The MMS collects field data on API gravity, which can be used to derive expressions to adjust the production price, but because the correlation between gravity and sulfur content is weak, introduction of a quality-adjusted hydrocarbon price does not add materially to the reliability of price forecasting. The adjustment is considered of secondary importance to the model results, and only prices derived from historic data are employed.

Abandonment Time

A structure is abandoned when the production and revenue profile forecast converges to the threshold level of the structure. Threshold levels vary with many factors such as the water depth, structure type, operator size, hydrocarbon production, number of structures on the lease, etc., but by examining the characteristics of structures that have been removed near the time of their abandonment, historical threshold limits serve as a guide and predictor of future levels (Table E.5). From Table E.5, observe that threshold levels generally increase with water depth and structure type, and are dependent on whether oil or gas is being produced. Economies of scale are also frequently present, so that threshold levels are slightly lower if more than one producing structure exists on a lease; e.g., structures that exist on leases with no other infrastructure at the time of removal (lease category I) exhibit an average annual production and revenue threshold of 57,000 BOE and $734,000; for leases with more than one producing structure at the time of removal (lease category II), the average production and revenue thresholds are 33,000 BOE and $423,000. For further details, refer to Chapter 4.

Decommissioning Cost

The total cost to decommission a structure is decomposed according to three categories – plugging and abandonment ($C_1(s)$), structure removal ($C_2(s)$), and site clearance and verification ($C_3(s)$). Functional expressions for each cost component have previously been derived based on market data sampling. Representative functions follow:

$$C_1(s) = \$150,000/\text{well},$$

$$C_2(s) = \$506.9 + 579.1ST - 0.24WD - 262.5REEF + 53.1CF,$$

$$C_3(s) = \$1,061 + 66WD + 17,752ST + 405AGE + 8,919G,$$

where WD = water depth (in feet), ST = structure type ($ST = 0$, caisson; $ST = 1$, otherwise), $REEF$ = reefing option ($REEF = 0$, structure not reefed; $REEF = 1$, structure reefed), CF = complexity factor (unitless), AGE = age upon removal (in years), G = Gorilla net application ($G = 0$, Gorilla net not applied; $G = 1$, Gorilla net applied).

The complexity factor is defined as the total number of piles and wells associated with the structure, and $C_2(s)$ needs to be scaled by a factor of 1,000. The reefing option is considered a random variable that depends upon water depth and planning area. Application of the Gorilla net is assumed to occur in 1-in-4 site clearance and verification operations.

5.4. Model Results

5.4.1. Model Statistics: Offshore structures can be aggregated according to structure type, production type, water depth, planning area, block type, or any other suitable categorization. Let the number of structures removed in year t under category Γ be denoted as $NR(\Gamma, t)$. The time variable runs from the observation year, $t = \tau$, until the year the last structure in Γ is removed, $t = T$:

$$NR(\Gamma) = (NR(\Gamma, \tau), NR(\Gamma, \tau+1), \cdots, NR(\Gamma, T)).$$

$NR(\Gamma)$ is a stochastic process.

The present value to remove structures is denoted,

$$C(\Gamma, d) = \sum_{s \in \Gamma} \frac{C(s)}{(1+d)^{t_r(s)-\tau}}.$$

The total undiscounted ($d = 0$) cost to remove all structures in category Γ is constant.

5.4.2. Removal and Cost Forecast: The number of structures expected to be removed in the CGOM using the production and revenue threshold models is depicted in Figure E.1 (production threshold) and Figure E.2 (revenue threshold). The total cost to decommission the CGOM from 2005-2025 is estimated to range between $4.9 - 5.2 billion as shown in Figure E.3 and Figure

71

E.4. The net present value of the total cost using a 10% discount factor is computed to range between $2.3 – 2.5 billion.

The number of structures expected to be removed in the WGOM according to the production and revenue threshold models is depicted in Figure E.5. The total cost to decommission the WGOM from 2005-2020 is shown in Figure E.6. The net present value of the total cost using a 10% discount factor is computed to range between $367 – 378 million.

5.4.3. Model Discussion: Modeling removal processes require a number of structural assumptions and parameterizations for decline parameters, economic limits, and removal obligations. The model results are closely linked to the model assumptions and parameterization so that changes in either of these factors will impact the forecast results.

The production threshold model dominates the revenue threshold model, and is considered less robust since hydrocarbon price is not incorporated in the analysis. It is expected that a significant number of structures removed over the next few years will remain in inventory on inactive leases, and thus the apparent removal rate will be "smoothed" out over the near-term horizon.

Decommissioning cost patterns reflect the removal forecast and the relative magnitude of abandonment. Fixed platforms in deepwater are significantly more costly to remove than shallow water caissons, for instance, and this is reflected in the higher proportional cost for fixed platform removals.

5.5. Limitations of the Analysis

Production and Revenue Model

A reliable production forecast early in the life of a field can only be developed with knowledge of the development plan, reserve estimates and production capacity, and since estimates of these parameters are either unknown or uncertain, a large degree of uncertainty exists in forecasting production profiles for structures that have yet to reach peak production. During the mid-point in the life of a field, a different sort of uncertainty arises since the production profile and the drive mechanisms of the field are reasonably well understood, but the strategic decisions of the operator are unknown. Will the operator invest additional capital, seek a joint operating agreement, or divest the structure? Leases are held by a wide variety of working interest owners and are carved up over time and sold off or become subject to a variety of joint venture/farmout type arrangements. As properties change hands, the capital expenditures and operating cost profiles change, and subsequently, the structure of production profiles can be expected to change.

The collection of wells in the MMS database associated with a structure is not known completely, since not all wells maintain a structure identification code, and so to estimate structure production, a correspondence was required to identify unassigned wells with a given structure or set of structures. Production from wells with no structure identifiers are assigned based upon the criteria

$$w^u \leftrightarrow \{s \mid \min_{s \in I} d(w^u, s)\},$$

72

where $d(w^u, s)$ represents the distance from unassigned well w^u to structure $s \in l$. Model uncertainty is introduced into the forecast model since this well-structure assignment is arbitrary. Fortunately, the number of wells requiring assignment is reasonably small, and so the error associated with the assignment is believed to be reasonably small.

The quality of production is not considered a primary factor in the forecast model and was assumed uniform across the GOM. For a first-order model, this is considered reasonable.

Abandonment Time

The link between abandonment and production/revenue threshold levels is not perfect, but is believed to provide a reasonably good indicator when an active structure will cease production. The economic limit serves to proxy abandonment time, and as with all proxy measures, a degree of uncertainty is associated with the inference.

Removal Time

Delay is the general principle in decommissioning decision making, because

- Delay transfers liability and the cost of decommissioning to the future,

- Delay allows economies of scale to be utilized (i.e., remove all the structures on the lease at once) which will reduce the cost of the operation,

- Delay allows structures to be requalified for new or marginal development plans and enhance field development options, and

- Federal regulations specify that an offshore structure can be held offshore as long as the lease remains producing, increasing the removal options and opportunities available to the operator.

Operators may remove idle structures early if decommissioning economics are favorable, or as strategic objectives dictate, but the input parameters that influence decision-making are generally unobservable.

5.6. Conclusions

The purpose of a forecast model is to gain insight into the nature of the operation, the system structure, and the factors that influence the model results. In this chapter, removal processes were modeled using a "ground-up" approach employing constructive techniques and a suitable parameterization to re-create the economic and behavior conditions that operators apply in decision-making. The number of structures removed and cost statistics form the model output. The analytic structure of the model was described, and through a suitable parameterization, structure removal forecast and cost metrics estimated across each planning area categorization.

REFERENCES

Allen, F.H. and R.D. Seba. 2003. *Economics of Worldwide Petroleum Production.* Oil and Gas Consultants International (OGCI), Inc., Tulsa, OK.

Berndt, E.R. 1991. *The Practice of Econometrics.* Addison-Wesley, Reading, MA.

Brandon, J.W., B. Ramsey, J. Macfarlane, and D. Dearman. 2000. "Abrasive water-jet and diamond wire – cutting technologies used in the removal of marine structures," OTC 12002, Offshore Technology Conference, Houston, TX.

Brealey, R.A. and S.C. Myers. 1991. *Principles of Corporate Finance.* 4th Edition, McGraw-Hill, New York, NY.

Daniels, G.R. 1994. *Hurricane Andrew's Impact on Natural Gas and Oil Facilities on the Outer Continental Shelf: Interim Report as of November 1993.* U.S. Department of the Interior, Minerals Management Service, Engineering and Technology Division, Herndon, VA. OCS Report MMS 94-0031.

Dauterive, L. 2001. *Rigs-to-Reefs, Policy, Progress, and Perspective.* U.S. Department of the Interior, Minerals Management Service, Gulf of Mexico OCS Region, New Orleans, LA. OCS Report MMS 2000-073.

DeMarsh, G. 2000. "The use of explosives in decommissioning and salvage," OTC 12023, Offshore Technology Conference, Houston, TX.

Federal Register. 2002a. Department of the Interior, Minerals Management Service, 30 CFR Parts 250, 256, Oil and Gas and Sulphur Operations in the Outer Continental Shelf – Decommissioning Activities; Final Rule. 67(96):35398-35412.

Federal Register. 2002b. Taking and importing marine mammals; taking bottlenose dolphins and spotted dolphins incidental to oil and gas structure removal activities in the Gulf of Mexico. 67(148):49869-49875.

Gallun, R.A., C.J. Wright, L.M. Nichols, and J.W. Stevenson. 2001. *Fundamentals of Oil & Gas Accounting.* 4th Edition, PennWell Books, Tulsa, OK.

Gerwich, B.C., Jr. 2000. *Construction of Marine and Offshore Structures.* CRC Press, Boca Raton.

Gitschlag, G.R., M.J. Schirripa, and J.E. Powers. 2000. *Estimation of fisheries impacts due to underwater explosives used to sever and salvage oil and gas platforms in the U.S. Gulf of Mexico.* U.S. Department of the Interior, Minerals Management Service, Gulf of Mexico OCS Region, New Orleans, LA, OSC Study MMS 2000-087.

Graff, W.J. 1981. *Introduction to Offshore Structures.* Gulf Publishing Company, Houston, TX.

Greca, A.D. 1996. Offshore facility removal: How to save cost and marine resources. *SPE European Petroleum Conference*, Milan, Italy, Oct. 22-24.

Griffen, W.S. 1998. "Evaluation of the global decommissioning regulatory regime," OTC 51149, Offshore Technology Conference, Houston, TX.

Guegel, A. 2001. "Cutting to the chase in Gulf." *Upstream News.* December 7, p.25.

Hakam, A. and W. Thornton. 2000. "Case history: Decommissioning, reefing, and reuse of Gulf of Mexico platform complex," OTC 12021, Offshore Technology Conference, Houston, TX.

Hyne, N.J. 1995. *Nontechnical Guide to Petroleum Geology, Exploration, Drilling, and Production.* PennWell Books, Tulsa, OK.

Johnston, D. 2000. "Current developments in production sharing contracts and international petroleum concerns: Economic modeling/auditing: Art or science?" *Petroleum Accounting and Financial Management Journal.* 19(3):120-138.

Kaiser, M.J., D.V. Mesyanzhinov, and A.G. Pulsipher. 2002. "Explosive removal of offshore structures in the Gulf of Mexico." *Ocean and Coastal Management.* 45(8):459-483.

Kaiser, M.J., A.G. Pulsipher, and R.C. Byrd. 2003. "Decommissioning cost functions in the Gulf of Mexico." *ASCE Journal of Waterways, Ports, Harbors, and Ocean Engineering.* 129(6):1-11.

Kaiser, M.J. and A.G. Pulsipher. 2003. "The cost of explosive severance operations in the Gulf of Mexico." *Ocean and Coastal Management.* 46(6-7):701-740.

Kemp, A.G. and L. Stephen. 1997. "Economic and Fiscal Aspects of Decommissioning Offshore Structures," in *Decommissioning Offshore Structures* (D.G. Gorman and J. Neilson, Eds.), Springer Verlag, New York, 79-123.

Kirby, S. 1999. "Donan field decommissioning project," OTC 10832, Offshore Technology Conference, Houston, TX.

Manago, F. and B. Williamson (eds). 1998. *Proceedings: Public Workshop, Decommissioning and Removal of Oil and Gas Facilities Offshore California: Recent Experiences and Future Deepwater Challenges.* Coastal Research Center, Marine Science Institute, University of California, Santa Barbara, CA, OCS Study MMS 1998-0023.

McClelland, B. and M.D. Reifel (Ed.). 1986. *Planning and Design of Fixed Offshore Platforms.* Van Nostrand Reinhold Company, New York.

McKay, M., J. Nides, and D. Vigil, eds. 2002. Proceedings: *Gulf of Mexico Fish and Fisheries,* "*Bringing together new and recent research.* U.S. Dept. of the Interior, Minerals

Management Service, Gulf of Mexico OCS Region, New Orleans, LA, OCS Study MMS 2002-004.

Mian, M.A. 2002. *Project Economics and Decision Analysis, Vol. 1: Deterministic Models.* PennWell Books, Tulsa, OK.

National Research Council. 1985. *Disposal of Offshore Platforms.* National Academy Press, Washington, D.C.

National Research Council. 1986. *An Assessment of Techniques for Removing Offshore Structures*, National Academy Press, Washington, D.C.

Ness, S., E. Lind, and J. Sandgren. 1996. Methodology for assessing environmental impacts of offshore installation abandonment and disposal. *SPE International Conference on Health, Safety, and Environment*, New Orleans, LA, June 9-12.

O'Connor, P. 1998. Case studies of platform re-use in the Gulf of Mexico. *International Conference on the Re-Use of Offshore Production Facilities*, Netherlands, Oct. 13-14.

Pulsipher, A.G., ed. 1996. *Proceedings: An International Workshop on Offshore Lease Abandonment and Platform Disposal: Technology, Regulation, and Environmental Effects, New Orleans, LA, April 15-17.*

Pulsipher, A.G. and W.B. Daniel. 2000. "Onshore disposition of offshore oil and gas platforms: Western politics and international standards." *Ocean & Coastal Management.* 43:973-995.

Pulsipher, A.G., J.E. Kiesler, V. Mackey, and W. Daniel. 1996. "Explosives remain preferred method for platform abandonment." *Oil and Gas Journal.* 94(19):64-70.

Reggio, V. C. 1989. *Petroleum structures as artificial reefs: A compendium.* U.S. Department of the Interior, Minerals Management Service, Gulf of Mexico OCS Region, New Orleans, LA, OCS Study MMS 89-0021.

Roberts, S.P. and K. Hollingshead. 2000. *Marine mammal regulatory issues and the explosive removal of offshore structures: The small take authorization program.* NOAA Fisheries, Office of Protected Resources, Marine Mammal Conservation Division, Washington, D.C.

Roberts, T. and J. Mitchell. 1997. "North Sea decommissioning: Valuing the options." Oxford Institute for Energy Studies, Oxford, England, EE21.

Rose, P.R. 2001. "Risk analysis and management of petroleum exploration ventures," AAPG Methods in Exploration Series, No. 12, American Association of Petroleum Geologists, Tulsa, OK.

Seba, R. 2003. *Economics of Petroleum Exploration.* CGI-International, Houston, TX.

Shaw, M. 2000. "Garden banks 388 deepwater decommissioning: Regulatory considerations, issues and challenges," OTC 12123, Offshore Technology Conference, Houston, TX.

Studenmund, A.H. 2001. *Using Econometrics: A Practical Guide*. Fourth Edition, Addison Wesley Longman, Boston, MA.

Thompson, R.S. and J.D. Wright. 1984. *Oil Property Evaluation*. Thompson-Wright Associates, Golden, Colorado.

Thornton, W.L. 1989. "Case history: Salvage of multiple platforms and pipelines offshore Texas." OTC 6074, Houston, TX.

Thornton, W.L. 1995. "Project Management: A checklist." *Supplement to Petroleum Engineer International*. (July):12-13.

Thornton, W. and J. Wiseman. 2000. "Current trends and future technologies for the decommissioning of offshore platforms," OTC 12020, Offshore Technology Conference, Houston, TX.

APPENDIX A
CHAPTER 1 FIGURES AND TABLES

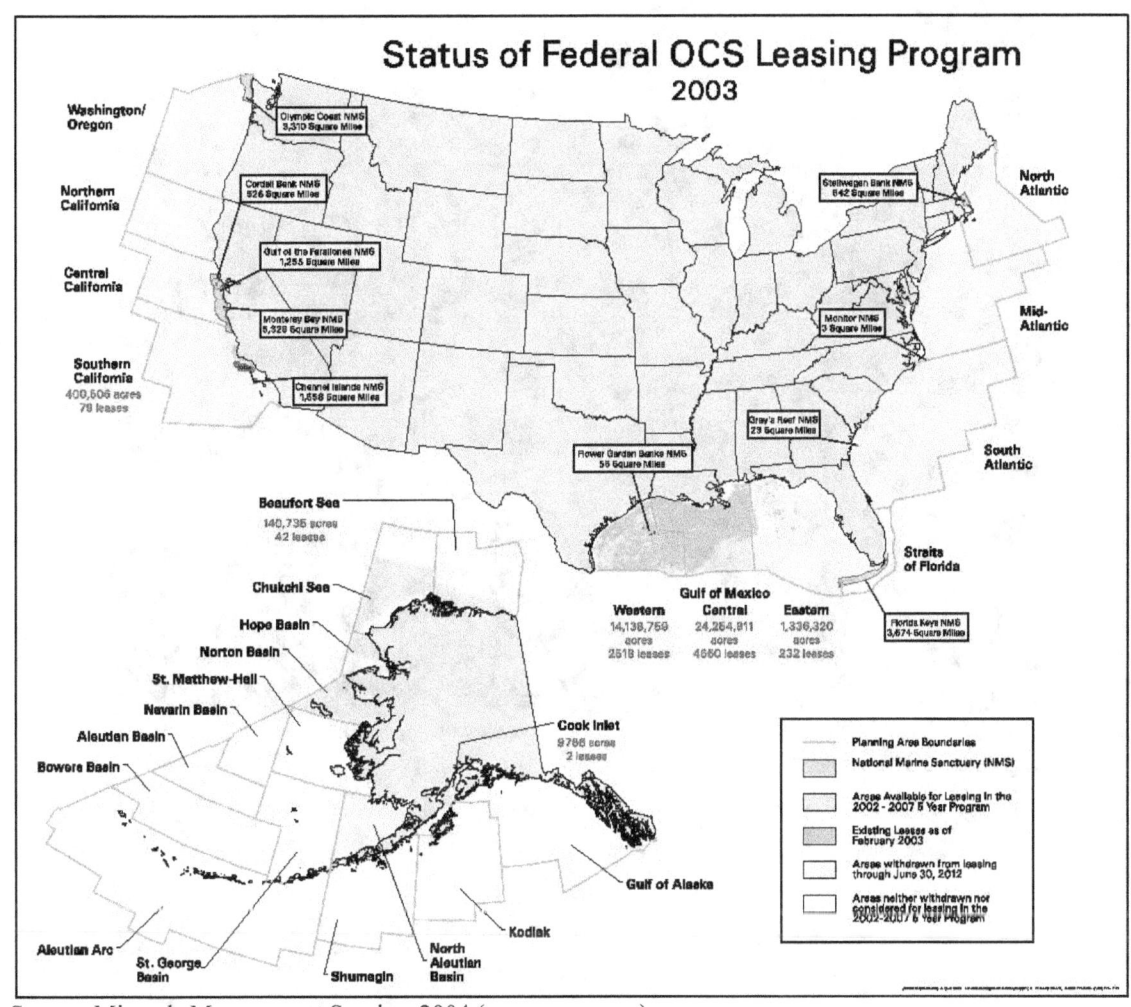

Source: Minerals Management Service, 2004 (www.mms.gov).

Figure A.1: Federal Outer Continental Shelf Leasing Program.

Source: Twachtman Synder & Byrd, Inc., 2000 (www.tsboffshore.com).
Figure A.2: Caisson Structures.

Source: Twachtman Synder & Byrd, Inc., 2000 (www.tsboffshore.com).
Figure A.3: Well Protector Structures.

Source: Twachtman Synder & Byrd, Inc., 2000 (www.tsboffshore.com).
Figure A.4: Fixed Platform Structures.

Table A.1

Total Number of Structures Installed and Removed by Water Depth and Planning Area in the Gulf of Mexico (1947-2001)

Water Depth Range (ft)	Installed			Removed		
	WGOM	CGOM	GOM	WGOM	CGOM	GOM
0-10	2	103	105	1	37	38
11-20	0	527	527	0	263	263
21-30	2	695	697	1	291	292
31-40	20	660	680	5	190	195
41-50	67	597	664	33	216	249
51-75	216	834	1,050	84	305	389
76-100	123	439	562	40	140	180
101-125	50	282	332	20	86	106
126-150	52	242	294	20	64	84
151-175	48	170	218	19	37	56
176-200	51	190	241	19	45	64
Subtotal	631	4,739	5,370	242	1,674	1,916
201-656	123	447	570	22	63	85
657-2,624	14	19	33	2	1	3
2,624+	2	6	8	0	0	0
Subtotal	139	472	611	24	64	88
Total	770	5,211	5,981	266	1,738	2,004

Footnote: Structures are defined to include all caissons, well-protectors, fixed platforms, and floating configurations located within the federal offshore waters of the Gulf of Mexico.

Table A.2

Average Annual Number of Structures Installed and Removed in the Gulf of Mexico According to Water Depth and Planning Area (1996-2001)

Water Depth Range (ft)	Installed			Removed		
	WGOM	CGOM	GOM	WGOM	CGOM	GOM
0-10	(0, 0)	(2.8, 2.5)	(2.8, 2.3)	(0.2, 0.4)	(2.6, 2.7)	(2.8, 2.8)
11-20	(0, 0)	(6, 2.9)	(6, 2.9)	(0, 0)	(16.2, 8.6)	(16.2, 8.6)
21-30	(0.2, 0.4)	(11.2, 5.2)	(11.4, 5.4)	(0, 0)	(11.8, 6.6)	(11.8, 6.6)
30-40	(0.8, 0.4)	(13.4, 4.0)	(14.4, 3.6)	(0.6, 1.3)	(10, 5.2)	(10.6, 6.2)
41-50	(2.8, 1.9)	(5.8, 1.3)	(8.6, 2.2)	(2.6, 1.8)	(14.6, 7.7)	(17.2, 8.5)
51-75	(7.6, 1.9)	(18, 6.4)	(25.6, 6.0)	(7.8, 6.5)	(19.8, 6.4)	(26.7, 10.9)
76-100	(2.4, 1.7)	(11, 4.3)	(13.4, 4.8)	(4.6, 4.4)	(8, 2.2)	(12.6, 3.1)
101-125	(1, 0.7)	(8.2, 4.1)	(9.2, 4.8)	(2, 2)	(6.8, 7.4)	(8.8, 9.2)
126-150	(1.4, 1.1)	(6.6, 2.5)	(8, 2.9)	(0.8, 1.3)	(4.2, 1.5)	(5, 2.1)
151-175	(0.8, 0.8)	(5.2, 3.3)	(6, 3.1)	(1.2, 0.8)	(2.8, 0.8)	(4, 0)
176-200	(1, 1.2)	(5, 2.6)	(6, 3.8)	(0.8, 0.8)	(3.8, 1.5)	(4.6, 1.8)
Subtotal	(18, 3.8)	(93, 12.6)	(111, 13.2)	(20.6, 18.6)	(100.6, 17.8)	(121.2, 21.3)
201-656	(3.6, 1.5)	(16.2, 4.9)	(19.8, 6.3)	(1.4, 1.1)	(5.4, 2.6)	(6.8, 3.3)
657-2,624	(1.2, 0.4)	(1.4, 1.5)	(2.6, 1.9)	(04, 0.5)	(0, 0)	(0.4, 0.5)
2,624+	(0.2, 0.4)	(1.2, 0.8)	(1.4, 1.1)	(0, 0)	(0, 0)	(0, 0)
Subtotal	(5, 1.6)	(18.8, 5.2)	(21.6, 7.6)	(1.8, 1.2)	(5.4, 2.6)	(7.2, 3.3)
Total	(23, 4.1)	(111.8, 13.7)	(134.8, 14.8)	(22.4, 8.7)	(106.0, 17.9)	(128.4, 21.6)

Footnote: The data entries are denoted by the coordinate pair μ, σ where μ represents the mean and σ the standard deviation of the structure data per water depth and planning area category

Table A.3

Average Annual Number of Structures Installed and Removed in the Gulf of Mexico According to Water Depth and Planning Area (1991-2001)

Water Depth Range (ft)	Installed			Removed		
	WGOM	CGOM	GOM	WGOM	CGOM	GOM
0-10	(0.1, 0.3)	(2.4, 2.3)	(2.5, 2.2)	(0.1, 0.3)	(2.2, 2.2)	(2.3, 2.3)
11-20	(0, 0)	(7.2, 3.2)	(7.2, 3.2)	(0, 0)	(17, 9.2)	(17, 9.2)
21-30	(0.1, 0.3)	(10.9, 4.3)	(11, 4.4)	(0, 0)	(14.6, 8.5)	(14.6, 8.5)
30-40	(0.6, 0.5)	(10.3, 4.8)	(10.9, 4.9)	(0.3, 0.9)	(10.5, 3.7)	(10.8, 4.3)
41-50	(1.8, 1.8)	(7.9, 3.5)	(9.7, 3.2)	(1.9, 1.4)	(13, 7.2)	(14.9, 7.7)
51-75	(8, 3.9)	(19.3, 7.1)	(27.3, 5.8)	(6.7, 5.3)	(20.1, 9.6)	(26.8, 12.8)
76-100	(3.9, 3.2)	(12.3, 4.3)	(16.2, 6.4)	(3.6, 3.3)	(8.6, 3.8)	(12.2, 4.7)
101-125	(1.4, 1.0)	(8.8, 3.8)	(10.2, 4.5)	(1.6, 1.6)	(5.5, 6.1)	(7.1, 7.5)
126-150	(1.3, 1.1)	(6.0, 2.6)	(7.3, 3.1)	(1.4, 1.8)	(4.8, 4.7)	(6.2, 5.2)
151-175	(1.6, 1.6)	(5.1, 3.3)	(6.7, 3.1)	(1.5, 1.8)	(2.7, 2.5)	(4.2, 2.8)
176-200	(0.8, 0.9)	(4.8, 2.5)	(5.6, 3.2)	(1.1, 0.3)	(3.6, 1.75)	(4.7, 2.3)
Subtotal	(19.6, 5.9)	(95, 13.3)	(114.6, 14.2)	(18.2, 7.3)	(102.6, 20.1)	(120.8, 22.8)
201-656	(3.5, 1.9)	(15.2, 5.8)	(18.7, 7.4)	(1.6, 1.3)	(5, 2.1)	(6.6, 2.7)
657-2,624	(0.9, 0.6)	(1.2, 1.2)	(2.1, 1.6)	(0.2, 0.4)	(0, 0)	(0.2, 0.4)
2,624+	(0.2, 0.4)	(0.6, 0.8)	(0.8, 1.0)	(0, 0)	(0, 0)	(0, 0)
Subtotal	(4.6, 2.0)	(17, 6.0)	(21.6, 7.6)	(1.8, 1.4)	(5, 2.1)	(6.8, 2.7)
Total	(24.2, 6.2)	(112, 14.6)	(136.3, 16.1)	(20, 7.4)	(107.6, 20.2)	(127.6, 23.0)

Footnote: The data entries are denoted by the coordinate pair μ, σ where μ represents the mean and σ the standard deviation of the structure data per water depth and planning area category

Table A.4

The Age Distribution of Active Structures in Shallow Water (0-60m) by Configuration Type and Planning Area (1947-2001)

Water Depth Range (m)	Age Distribution (Year)	Caisson			Well Protector			Fixed Platform		
		W	C	G	W	C	G	W	C	G
0-60	40⁺	0	8	7	0	20	18	<1	11	9
	31-40	<1	13	12	11	19	19	4	23	20
	21-30	<1	16	15	13	22	21	23	23	23
	11-20	33	30	30	40	21	23	47	21	25
	1-10	63	33	33	36	17	19	25	23	23
Total	Total	79	1,073	1,152	47	355	402	252	1,617	1,800

Footnote: W, C, G denote WGOM, CGOM, and GOM.

Table A.5

Average Age of Structures Upon Removal by Water Depth, Configuration Type, and Planning Area (1947-2001)

Water Depth Range (m)	Caisson		Well Protector		Fixed Platform	
	W	C	W	C	W	C
0-60	(7.1, 4.7)	(15.9, 10)	(16.3, 1.3)	(17.4,10.2)	(9.3, 5.3)	(16.9, 10.8)
61-200			(9.5, 1)	(18, 13.5)	(10.3, 6.4)	(12.6, 8.1)

Footnote: The data entries are denoted by the coordinate pair (μ, σ), where μ represents the mean and σ is the standard deviation of the structure data per water depth and planning area category $\Gamma_{i,j}$. W, C, G denote WGOM, CGOM, and GOM.

Table A.6

Number of Structures Removed (*R*), Structures Removed by Explosive Technique (*R_E*), and the Percentage of Explosive Removals (*p_E*) as a Function of Water Depth and Planning Area (1986-2001)

Water Depth Range (ft)	WGOM			CGOM			GOM		
	R	R_E	p_E	R	R_E	p_E	R	R_E	p_E
0-10	1	1	100	20	11	55	21	12	54
11-20				210	71	34	210	71	34
21-30	1	1	100	208	145	70	209	146	70
31-40	4	3	75	150	88	59	159	91	59
41-50	31	22	71	155	107	69	186	129	69
51-75	78	34	44	238	130	55	316	164	52
76-100	41	20	71	109	61	56	150	81	54
101-125	19	12	63	81	45	56	100	57	57
126-150	20	15	75	60	49	82	80	64	80
151-175	17	10	59	34	23	68	51	33	65
176-200	16	13	81	44	26	59	60	39	65
201-656	23	17	74	64	49	77	87	66	76
657-2,624	2	1	50				2	1	50
2,624$^+$									
Water Depth Range (m)	WGOM			CGOM			GOM		
	R	R_E	p_E	R	R_E	p_E	R	R_E	p_E
0-60	228	131	57	1,309	756	58	1,537	887	58
61-200	23	17	74	64	49	77	87	66	76
200$^+$	2	1	50				2	1	50
Total	253	149	58	1,373	805	58	1,626	954	59

89

Table A.7

Number of Structures Removed (*R*), Structures Removed by Explosive Technique (*R_E*), and the Percentage of Explosive Removals (*p_E*) as a Function of Water Depth and Configuration Type for the Gulf of Mexico (1986-2001)

Water Depth Range (ft)	Caisson			Well Protector			Fixed Platform			All		
	R	R_E	p_E	R	R_E	p_E	R	R_E	p_E	R	R_E	p_E
0-10	14	5	36	1	1	100	6	6	100	21	12	57
11-20	170	55	32	11	4	36	29	12	41	210	71	34
21-30	137	98	71	23	12	52	49	36	73	209	146	70
31-40	89	49	55	12	6	50	53	36	68	154	91	59
41-50	99	71	72	28	22	79	59	36	61	186	129	69
51-75	141	63	44	48	32	67	127	69	54	316	164	52
76-100	51	19	37	25	12	48	74	50	68	150	81	54
101-125	25	6	24	14	7	50	61	44	72	100	57	57
126-150	9	6	67	12	10	83	59	48	81	80	64	80
151-175	8	5	63	8	7	88	37	22	59	51	33	65
176-200				11	6	55	41	28	68	60	39	65
201-656				6	5	83	81	61	75	87	66	76
657-2,624										2	1	50
2,624⁺												
Water Depth Range (m)	**Caisson**			**Well Protector**			**Fixed Platform**			**All**		
	R	R_E	p_E	R	R_E	p_E	R	R_E	p_E	R	R_E	p_E
0-60	749	381	51	193	119	62	595	387	65	1,537	887	58
61-200				6	5	83	81	61	75	87	66	76
200⁺										2	1	50
Total	749	381	51	199	124	62	676	448	66	1,626	954	59

Table A.8

Number of Structures Removed (R), Structures Removed by Explosive Technique (R_E), and the Percentage of Explosive Removals (p_E) as a Function of Time and Configuration Type for the Gulf of Mexico (1986-2001)

Year	Caisson			Well Protector			Fixed Platform			Total		
	R	R_E	p_E (%)	R	R_E	p_E (%)	R	R_E	p_E (%)	R	R_E	p_E (%)
1986				1	0	0	1	0	0	2	0	0
1987	10	0	10	2	0	0	7	0	0	19	0	0
1988	46	5	11	9	2	22	36	19	53	91	26	29
1989	46	34	74	7	6	86	34	30	88	87	70	80
1990	53	26	49	9	5	56	36	29	81	98	60	61
1991	54	26	48	16	11	69	44	36	82	114	73	64
1992	44	19	43	13	9	99	40	33	83	97	61	63
1993	77	49	64	30	12	40	61	41	67	168	102	61
1994	42	22	52	16	14	88	66	51	77	124	87	70
1995	59	40	68	9	7	78	49	34	69	117	81	69
1996	48	13	27	15	8	53	56	29	52	119	50	42
1997	92	54	59	14	11	79	71	38	54	177	63	58
1998	35	14	40	11	8	73	29	13	45	75	35	47
1999	72	35	49	17	9	53	45	32	71	134	76	57
2000	49	37	76	19	13	68	66	42	64	134	92	69
2001	22	7	32	11	9	82	35	21	60	68	37	54
Total	749	381	51	199	124	62	676	448	66	1,624	953	59

Table A.9

Number of Structures Removed (R) and Percentage of Structures Removed (p) Grouped According to Age Upon Removal and Planning Area (1986-2001)

Age	Number of Structures (R)			Percentage p (%)		
	WGOM	CGOM	GOM	WGOM	CGOM	GOM
0-10	159	479	638	63	35	39
11-20	72	405	477	28	29	29
21-30	13	282	295	5	21	18
30^+	9	207	216	4	15	13
Total	253	1,373	1,626	100	100	100

Footnote: $p = R/R_T$, where R_T denotes the total number of structures per planning area.

Table A.10

Number of Structures Removed Using Explosives (R_E) and the Percentage of All Structures Removed Using Explosives (p_E) Grouped According to Age Upon Removal and Planning Area (1986-2001)

Age	Number of Structures (R_E)			Percentage p_E (%)		
	WGOM	CGOM	GOM	WGOM	CGOM	GOM
0-10	82	229	311	52	48	49
11-20	50	253	303	69	62	64
21-30	9	172	181	69	61	61
30^+	8	151	159	89	73	74
Total	149	805	954	59	59	59

Footnote: $p = R_E/R$, where the R value are obtained from Table A.8.

Table A.11

Number of Structures Removed (*R*) and Percentage of Structures Removed (*p*) as a Function of Water Depth and Age Upon Removal (1986-2001)

Water Depth Range (m)	Age Upon Removal (yr)				
	0-11	11-20	21-30	30^+	Total
0-60	600	441	283	213	1,537
61-200	36	36	12	3	87
200^+	2	0	0	0	2
Total	638	477	295	216	1,626
Water Depth Range (m)	Percentage *p* (%)				
	0-11	11-20	21-30	30^+	Total
0-60	39	39	18	14	100
61-200	41	41	14	3	100
200^+	100	-	-	-	100
Total	638	477	295	216	100

Footnote: $p = R/R_T$, where R_T denotes the total number of structures per planning area.

Table A.12

Number of Structures Removed Using Explosives Techniques and the Percentage of Explosives Removal as a Function of Water Depth and Age Upon Removal (1986-2001)

Water Depth Range (m)	Age Upon Removal (yr)				
	0-11	11-20	21-30	30^+	Total
0-60	283	276	173	156	888
61-200	28	27	8	3	66
200^+	1	0	0	0	1
Water Depth Range (m)	Percentage p_E (%)				
	0-11	11-20	21-30	30^+	Total
0-60	47	63	61	73	58
61-200	78	75	67	100	76
200^+	50	-	-	-	50

Footnote: $p_E = R_E/R$, where the *R* values are obtained from Table A.8.

93

Table A.13

Number of Structures Removed (R), Number of Structures Removed by Explosive Technique (R_E), and the Percentage of Structures Removed by Explosives (p_E) Categorized According to Age and Configuration Type (1986-2001)

Age	Caisson			Well Protector			Fixed Platform		
	R	R_E	p_E (%)	R	R_E	p_E (%)	R	R_E	p_E (%)
0-10	295	116	59	75	40	53	266	154	58
11-20	204	115	56	52	35	67	221	153	69
21-30	157	83	53	36	21	58	102	77	75
30^+	93	67	72	36	28	78	87	64	74
Total	749	381	51	199	124	62	676	448	66

Table A.14

Percentage of Structures Removed by Configuration Type and Water Depth in the Gulf of Mexico (1986-2001)

Water Depth Range (m)	Caisson				
	0-11	11-20	21-30	30^+	Total
0-60	39	56	53	72	51
61-200					

Water Depth Range (m)	Well Protector				
	0-11	11-20	21-30	30^+	Total
0-60	52	67	58	77	62
61-200	100	67		100	83

Water Depth Range (m)	Fixed Platform				
	0-11	11-20	21-30	30^+	Total
0-60	55	68	77	73	65
61-200	76	76	67	100	75

Table A.15

Medium-Term Forecast of the Number of Structures Removed in the Gulf of Mexico by Explosive Technique (Model I)

Water Depth Range (m)	Forecast Horizon	CAIS		WP		FP		Total	
		W	C	W	C	W	C	W	C
0-60	2002-2006	14	97	4	64	52	155	70	316
	2007-2011	18	133	4	57	35	255	57	445
	2012-2016	8	104	5	41	62	222	75	367
	2017-2021	0	98	7	30	5	215	22	343
	2022-2026	0	115	9	29	0	203	9	347
Subtotal		40	547	29	221	165	1,052	234	1,818
61-200	2002-2006	0	0	5	0	19	62	24	62
	2007-2011	1	0	2	0	31	65	35	65
	2012-2016	1	1	0	0	15	83	16	84
	2017-2021	0	1	0	0	3	53	3	54
	2022-2026	0	2	0	17	0	17	0	36
Subtotal		2	4	7	17	69	278	78	301

Footnote: W, C denote WGOM and CGOM.

Table A.16

Medium-Term Forecast of the Number of Structures Removed in the Gulf of Mexico by Explosive Technique (Model II)

Water Depth Range (m)	Forecast Horizon	CAIS		WP		FP		Total	
		W	C	W	C	W	C	W	C
0-60	2002-2006	30	168	10	91	89	250	129	509
	2007-2011	10	221	11	83	62	317	83	621
	2012-2016	0	132	8	47	14	280	22	459
	2017-2021	0	26	0	0	0	205	0	231
Subtotal		40	547	29	221	165	1,052	234	1,820
61-200	2002-2006	1	0	4	0	48	107	53	104
	2007-2011	1	0	3	9	26	97	30	106
	2012-2016	0	2	0	8	0	77	0	87
	2017-2021	0	2	0	0	0	0	0	2
Subtotal		2	4	7	17	69	278	83	299

Footnote: W, C denote WGOM and CGOM.

APPENDIX B
CHAPTER 2 FIGURES AND TABLES

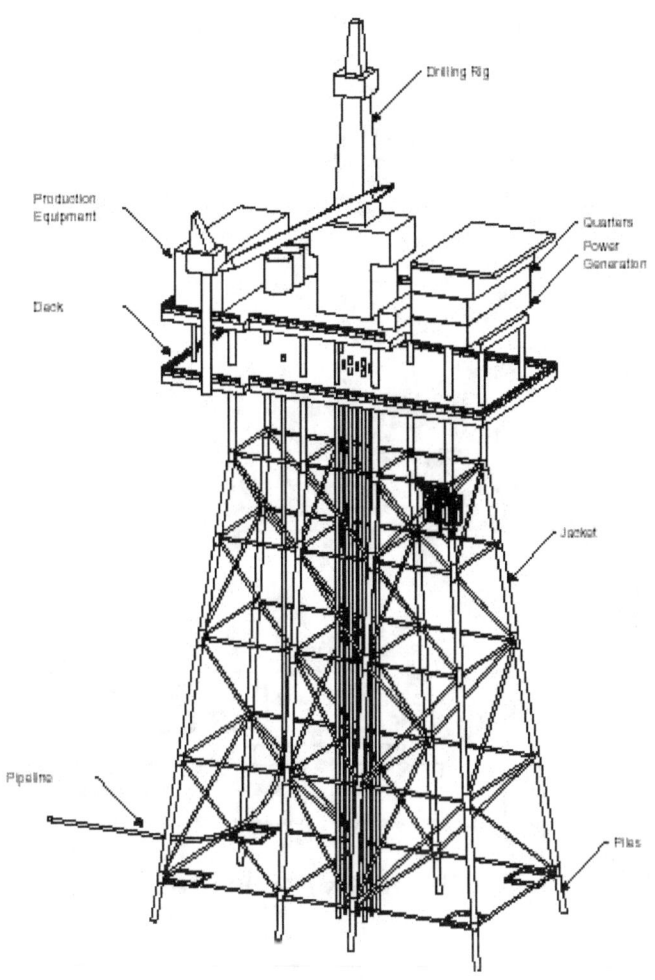

Figure B.1: Conventionally Piled Platform with Wells.

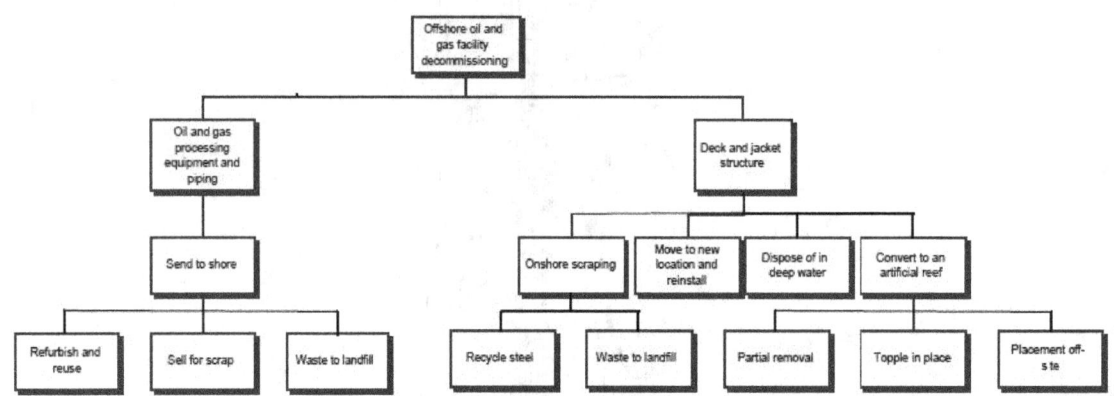

Figure B.2: Offshore Oil and Gas Facility Decommissioning Tree.

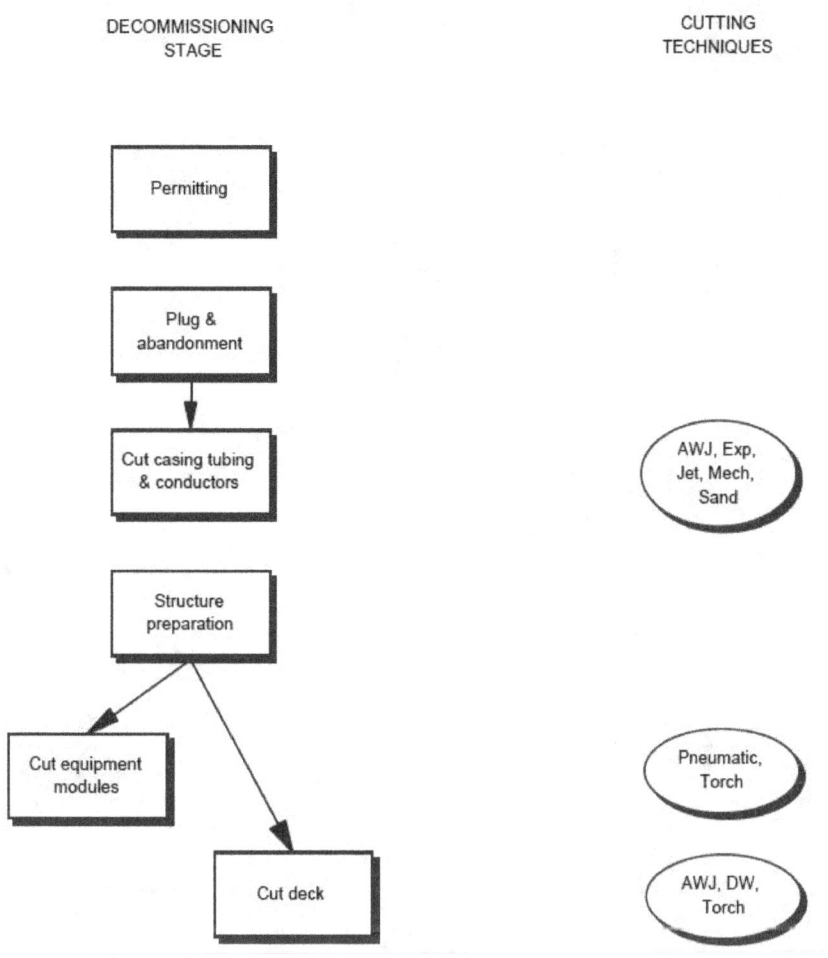

Figure B.3: Decommissioning Is Often a Severing Intensive Operation.

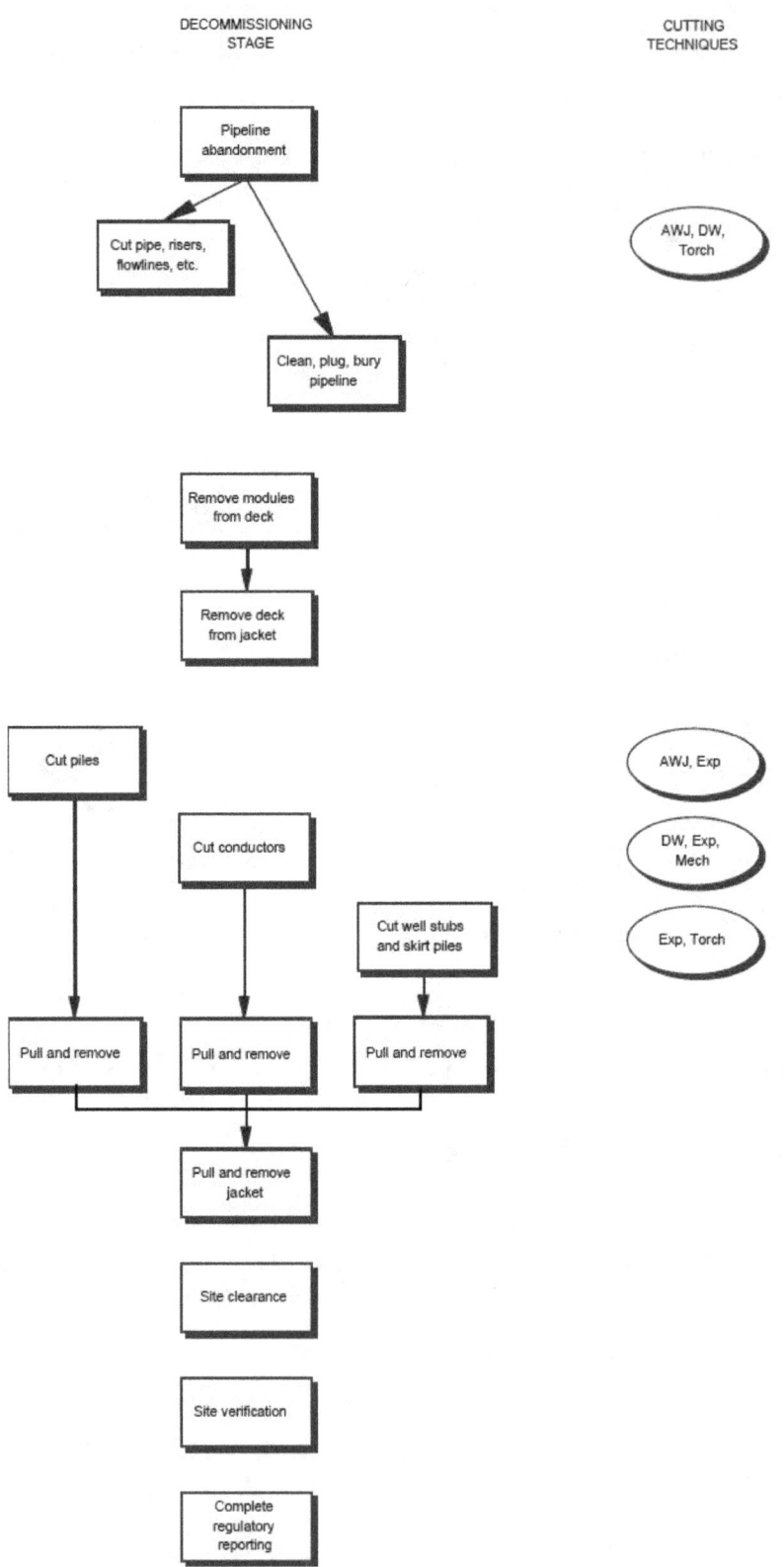

Figure B.3: Decommissioning Is Often a Severing Intensive Operation (continued).

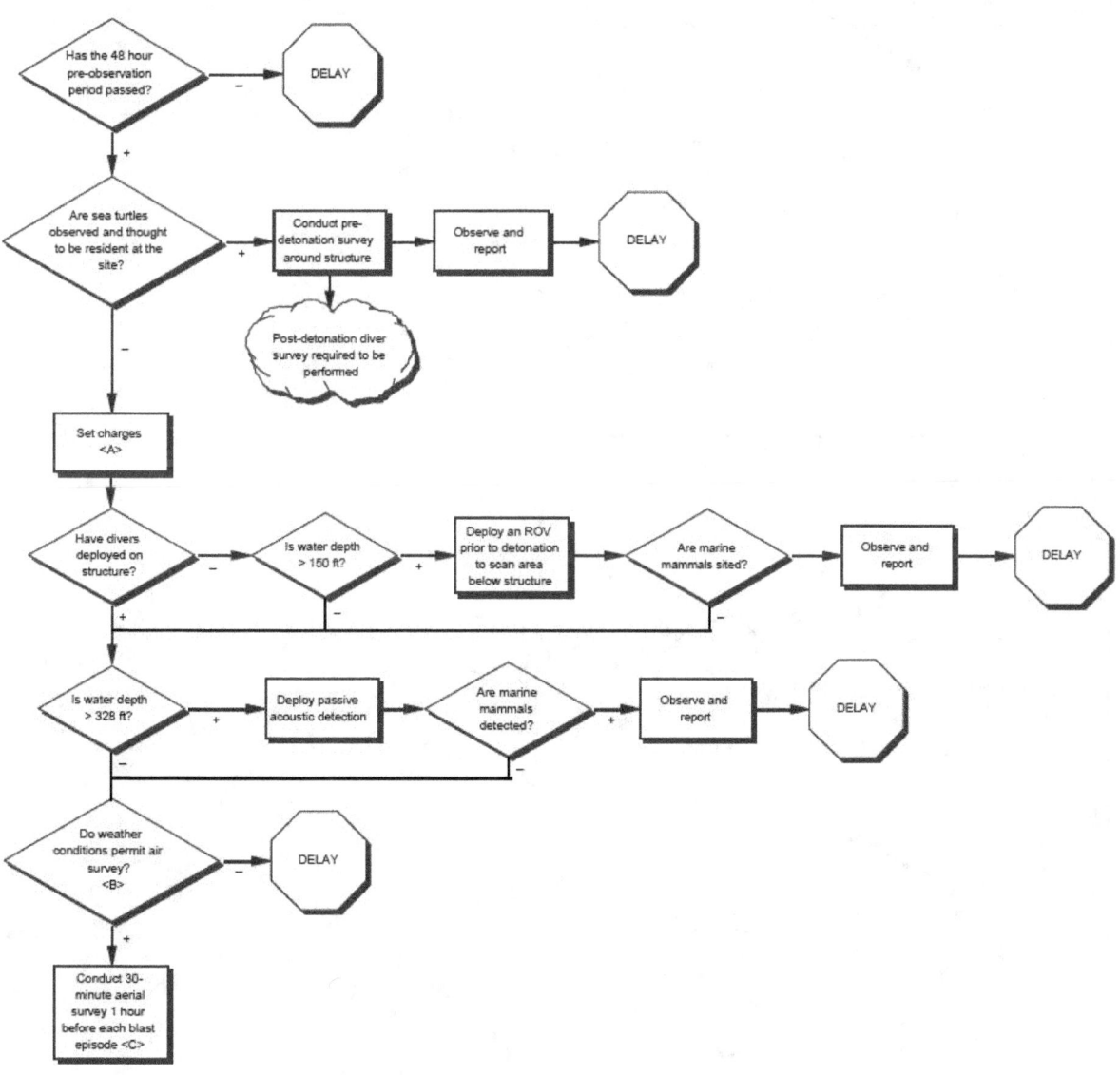

Figure B.4: The NMFS Observer Program (2002).

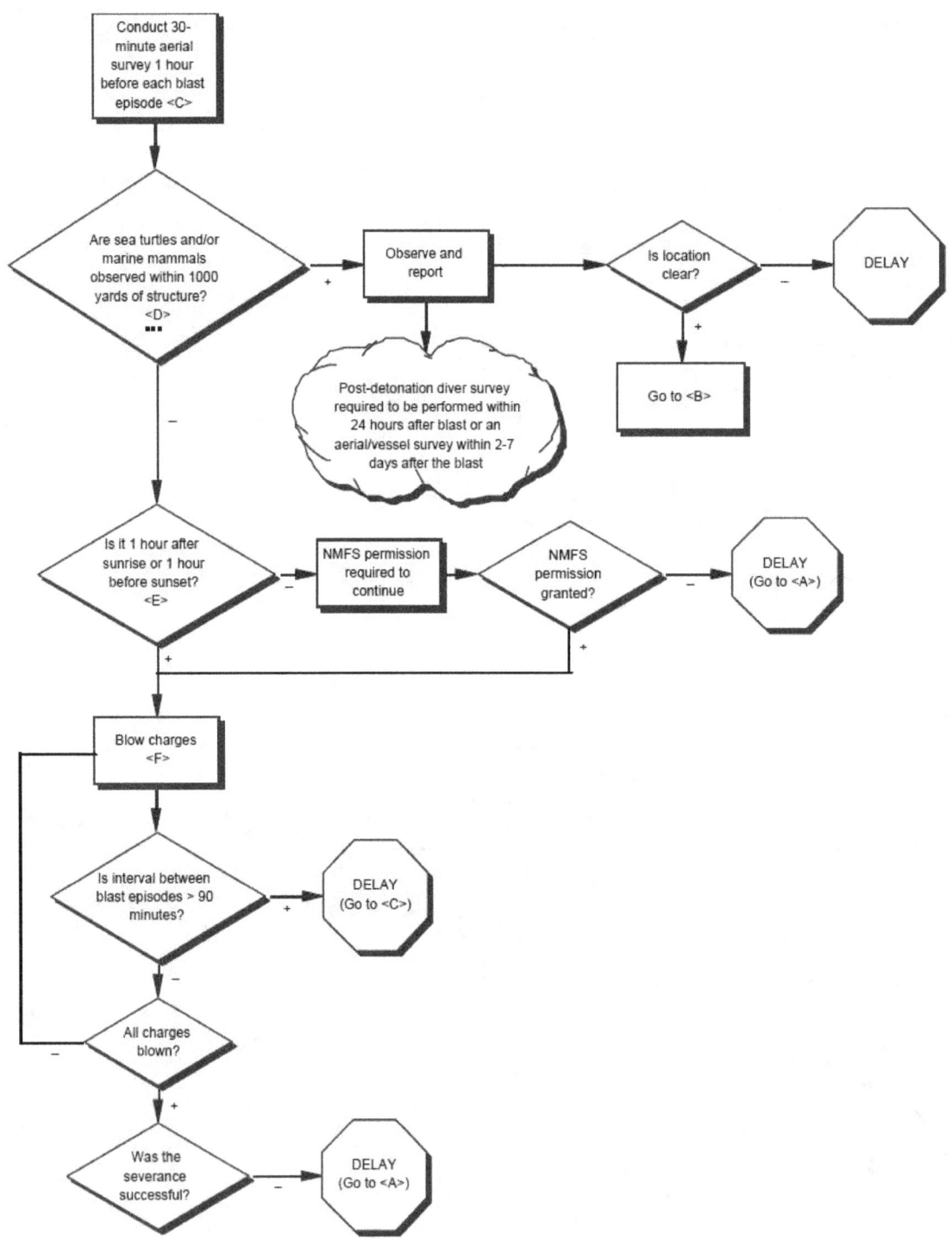

Figure B.4: The NMFS Observer Program (2002) (continued).

104

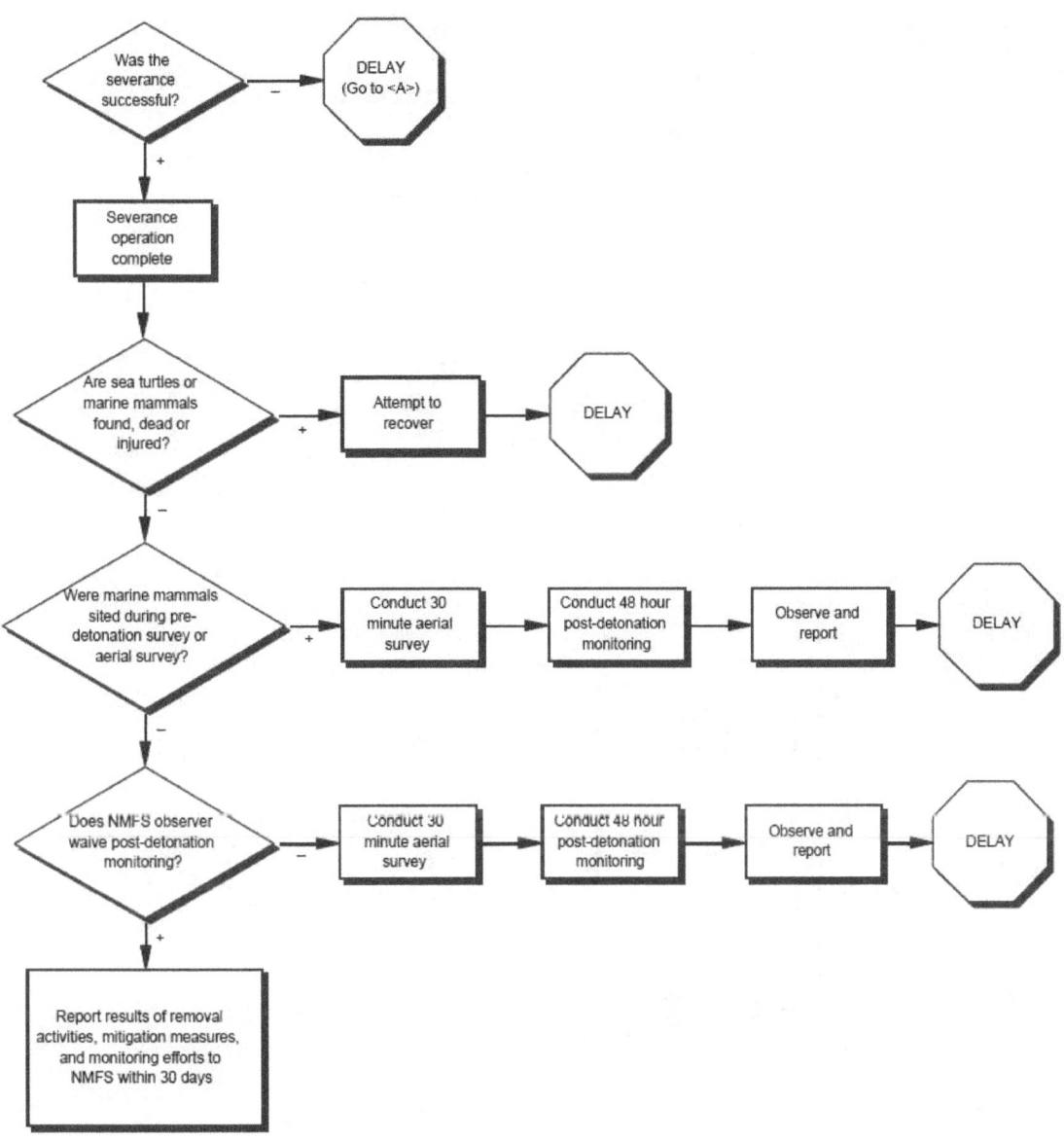

Figure B.4: The NMFS Observer Program (2002) (continued).

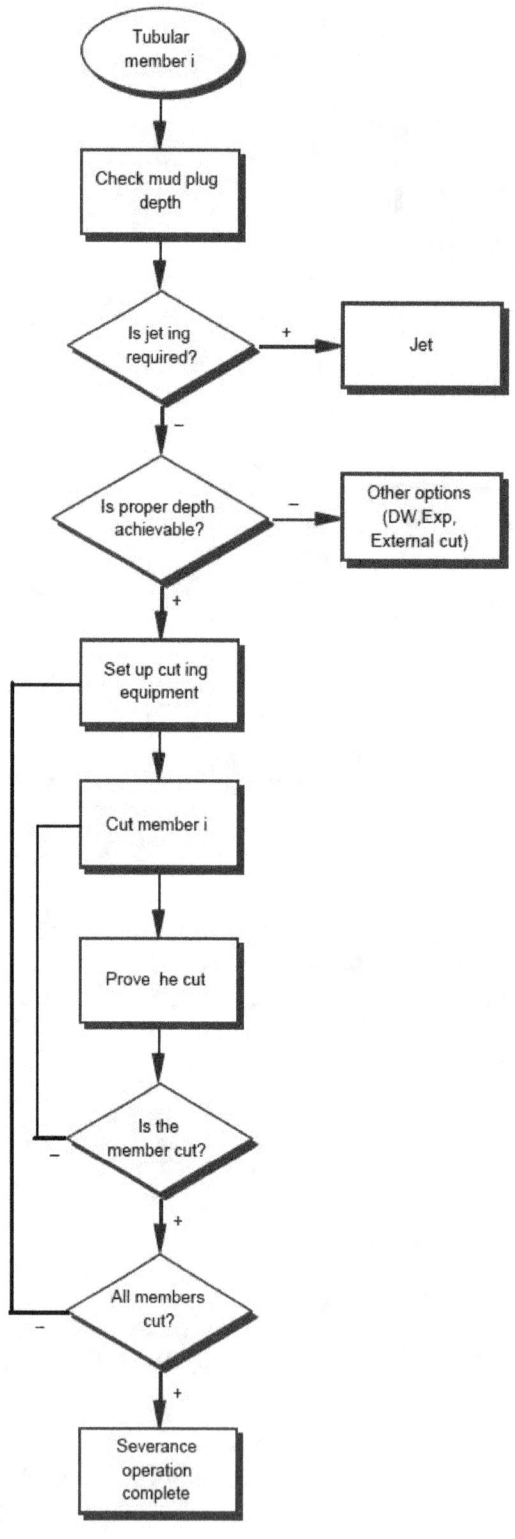

Figure B.5: The Abrasive Water Jet Cutting Process.

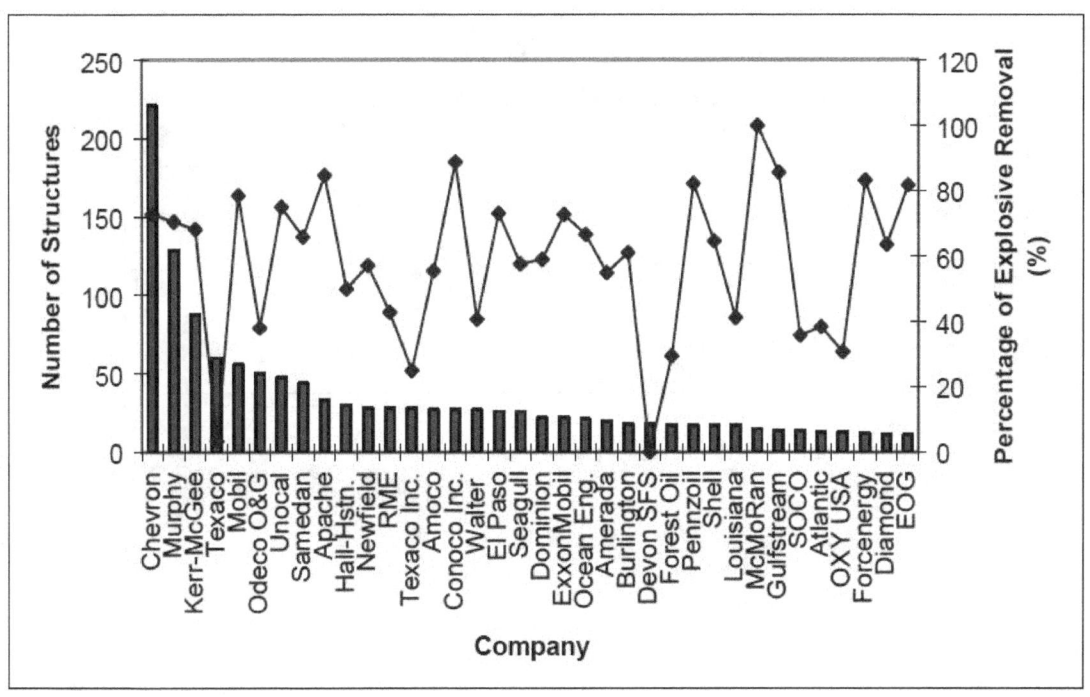

Figure B.6: Number of Structures Removed and Percentage of Explosive Removals by the "Top 36" Companies (1986-2001).

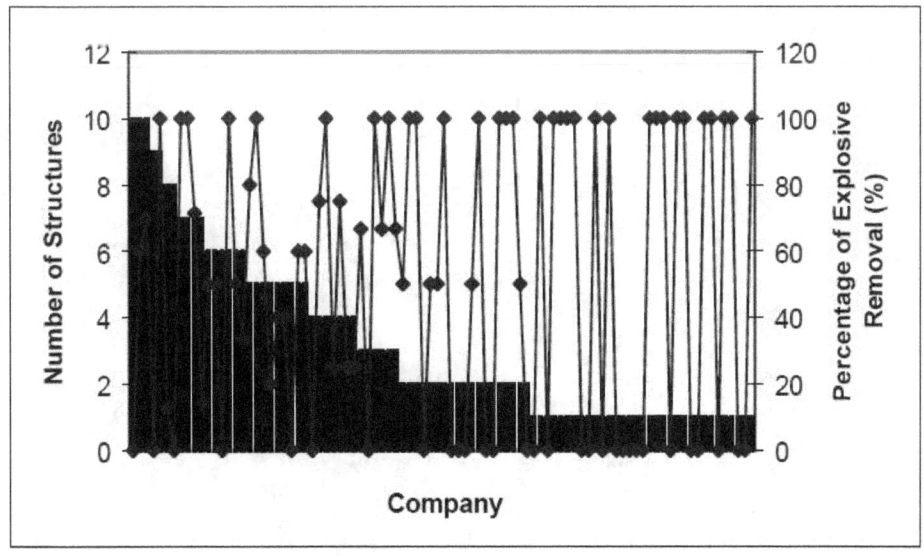

Figure B.7: Number of Structures Removed and Percentage of Explosive Removals by the "Bottom 91" Companies (1986-2001).

Table B.1

Gulf of Mexico Active and Removed Structures by Configuration Type, Water Depth and Number of Slots (1947-2001)

Configuration Type	Water Depth (feet)	Number of Slots	Active	Removed
Caisson				
	0-80		1076	921
	80-200		117	112
	200+		5	1
Well Protector				
	0-80			
		0-6	271	193
		7-12	8	8
		12+	2	2
	80-200			
		0-6	97	74
		7-12	16	12
		12+	3	2
	200+			
		0-6	26	3
		7-12	2	2
		12+	1	3
Non-Major Fixed Platform				
	0-80			
		0-6	291	85
		7-12	5	0
		12+	0	0
	80-200			
		0-6	155	23
		7-12	7	0
		12+	2	1
	200+			
		0-6	57	6
		7-12	9	0
		12+	3	0
Major Fixed Platform				
	0-80			
		0-6	511	272
		7-12	132	58
		12+	85	13
	80-200			
		0-6	304	168
		7-12	228	80
		12+	178	33
	200+			
		0-6	95	33
		7-12	109	28
		12+	212	24
TOTAL			4,007	2,159

Table B.2

Number of Structures Removed (R), Structures Removed by Explosive Technique (R_E), and the Percentage of Explosive Removals (p_E) as a Function of Water Depth and Configuration Type for the Gulf of Mexico (1986-2001)

Water Depth Range (m	Caisson			Well Protector			Fixed Platform			All		
	R	R_E	p_E	R	R_E	p_E	R	R_E	p_E	R	R_E	p_E
0-60	749	381	51	193	119	62	595	387	65	1,537	887	58
61-200				6	5	83	81	61	75	87	66	76
200$^+$										2	1	50
Total	749	381	51	199	124	62	676	448	66	1,626	954	59

Table B.3

Percentage of Explosive Removals by Configuration Type and Age Upon Removal in the Gulf of Mexico (1986-2001)

Age Upon Removal	Caisson		Well Protector		Fixed Platform	
(Year)	0-60m	61-200m	0-60m	61-200m	0-60m	61-200m
0-10	39		52	100	55	76
11-30	55		64	67	71	76
30$^+$	72		77	100	73	100

Table B.4

A Summary of Operator Involvement in Gulf of Mexico Structure Removals (1986-2001)

Number of Operators	Number of Structures Removed	Percentage of Total Structure Removals (%)	Percentage of Structures Removed with Explosives (%)	Number of Operators that Use Explosives Exclusively	Contribution to Subcategory Total (%)
Top 12	$R \geq 28$	50	63	0	0
Middle 24	$10 < R < 28$	30	57	1	4
Bottom 91	$R \leq 10$	20	51	33	55

Table B.5

Probit and Logit Model Results

Factor	Probit		Logit	
	Coefficient	(Z-Statistic)	Coefficient	(Z-Statistic)
Constant	0.1372	(1.02)	0.0815	(0.97)
ST	0.0563	(3.21)	0.1121	(3.26)
AGE	0.0051	(5.42)	0.0167	(5.49)
WD	0.0005	(0.10)	0.00001	(0.11)
R_p	0.65		0.63	

APPENDIX C
CHAPTER 3 TABLES

Table C.1

Design Space for Models I and II

Parameter	Model Ia	Model Ib	Model IIa	Model IIb	Model IIc
RES	N(100000, 10000)	N(100000, 20000)			
$d(t)$	U(0.08, 0.13)	U(0.08, 0.13)	U(0.08, 0.13)	U(0.08, 0.13)	U(0.08, 0.13)
P			LN(25, 3)	LN(25, 3)	
$P(t)$					LN(25, 3)
GR			U(15000, 30000)	U(15000, 50000)	U(15000, 50000)

Table C.2

Model I and II Regression Results

	$A(\varphi) = \alpha_0 + \alpha_1 \bar{d} + \alpha_2 RES + \alpha_3 P + \alpha_4 \bar{P} + \alpha_5 \overline{GR}$, $\varphi = \{I, II\}$				
Coefficient	Ia	Ib	IIa	IIb	IIc
α_0	-90.8(-32)	-81.1(-32)	53.4(210)	50.8(105)	51.5(19)
α_1	362.2(23)	382.3(20)	-229.8(-158)	-208.9(-88)	-206.9(-77)
α_2	0.0007(33)	0.0006(47)			
α_3			0.37(53)	0.36(22)	
α_4					0.29(3)
α_5			-0.00042(-88)	-0.00034(-115)	-0.00034(-102)
Iterations	1,000	1,000	1,000	1,000	1,000
R^2	0.62	0.72	0.97	0.96	0.95

Table C.3

Design Space for Decommissioning Model III

Parameter	Model IIIa, b	Model IIIc	Model IIId
$d(t)$	U(0.08, 0.13)	U(0.08, 0.13)	U(0.08, 0.13)
$P(t)$	LN(25, 3)	LN(25, 3)	LN(25, 3)
RES	U(0.10, 0.20)	U(0.10, 0.20)	U(0.10, 0.20)
k	U(0.9, 0.13)	U(0.9, 0.13)	U(0.9, 0.13)
T	U(0.30, 0.50)	U(0.30, 0.50)	U(0.30, 0.50)
E	U(4000, 8000)	U(4000, 8000)	U(4000, 8000)
l	0,0	1	2

Table C.4

Model III Regression Results

	$A(\varphi) = \alpha_0 + \alpha_1\,\overline{d} + \alpha_2\,\overline{P} + \alpha_3 ROY + \alpha_4 k + \alpha_5 T + \alpha_6 E,\ \varphi = \{III\}$			
Coefficient	IIIa	IIIb	IIIc	IIId
α_0	59.2(36)	58.5(78)	51.3(26)	56.8(26)
α_1	-206.1(-127)	-208.8(-285)	-214.8(-113)	-212.9(-101)
α_2	0.17(3)	0.22(8)	0.53(7)	0.34(4)
α_3	-12.6(-16)	-11.9(-33)	-11.5(-12)	-10.7(-10)
α_4	-4.8(-24)	-5.3(-57)	-4.8(-20)	-5.2(-20)
α_5	-7.3(-18)	-7.6(-42)	-6.8(-14)	-6.6(-13)
α_6	-0.00071(-35)	-0.00073(-80)	-0.00076(-32)	-0.00071(-27)
Iterations	1,000	5,000	1,000	1,000
R^2	0.95	0.96	0.94	0.92

APPENDIX D
CHAPTER 4 TABLES

Table D.1

Summary Statistics for Structures Removed in the Gulf of Mexico

Lease Categorization	Parameters	Caisson	Well Protector	Fixed Platform	All
I	$\overline{Q}\ (t_{lp})$ (BOE)	50,973	91,584	52,608	57,238
	$\overline{R}\ (t_{lp})$ ($)	604,667	1,147,691	698,593	733,805
	$IDLE$ (yr)	$(2.6, 3.9)^a$	(3.2, 4.6)	(1.9, 2.6)	(2.3, 3.2)
	Q^*/RES	(0.43, 0.23)	(0.38, 0.18)	(0.38, 0.20)	(0.40, 0.20)
	RES (MMBOE)	1.02	2.06	4.20	3.10
	n	170	73	389	632
II	$\overline{Q}\ (t_{lp})$ (BOE)	32,000	30,700	36,174	32,798
	$\overline{R}\ (t_{lp})$ ($)	392,006	384,021	522,685	422,867
	$IDLE$ (yr)	(6.5, 6.1)	(8.0, 7.6)	(4.6, 4.8)	(6.3, 6.2)
	Q^*/RES	(0.34, 0.21)	(0.29, 0.19)	(0.26, 0.16)	(0.31, 0.20)
	RES (MMBOE)	1.71	3.00	6.83	3.21
	n	397	124	171	692
III	$\overline{Q}\ (t_{lp})$ (BOE)	36,693	25,588	39,606	35,035
	$\overline{R}\ (t_{lp})$ ($)	531,191	354,210	575,522	507,282
	$IDLE$ (yr)	(3.9, 4.2)	(4.1, 4.0)	(3.6, 4.6)	(3.8, 4.3)
	Q^*/RES	(0.41, 0.25)	(0.35, 0.19)	(0.35, 0.18)	(0.40, 0.23)
	RES (MMBOE)	1.58	3.88	6.16	3.53
	n	78	35	53	166
IV	$\overline{Q}\ (t_{lp})$ (BOE)	39,061	22,832	39,429	37,495
	$\overline{R}\ (t_{lp})$ ($)	528,159	353,362	538,132	512,209
	$IDLE$ (yr)	(8.9, 7.3)	(6.7, 5.5)	(7.1, 7.5)	(8.4, 7.2)
	Q^*/RES	(0.27, 0.17)	(0.19, 0.10)	(0.23, 0.13)	(0.26, 0.18)
	RES (MMBOE)	2.21	2.51	11.83	3.71
	n	224	30	46	300

Footnote: (a) Ordered pair (x, y) denotes mean x and standard deviation y.

Table D.2

Average Production Threshold Levels At/Near the Year of Last Production – Lease Category I

Structure Type	Water Depth (ft)	n	$\overline{Q}(t_{lp})$ (BOE)	$\overline{Q}(t_{lp}-1)$ (BOE)	$\overline{Q}(t_{lp}-2)$ (BOE)	$\overline{Q}(t_{lp}-3)$ (BOE)
Caisson	0-100	140	50,106	120,809	158,112	144,080
	101-200	30	53,196	205,617	259,381	165,009
Well Protector	0-100	34	51,749	140,063	211,940	230,572
	101-200[a]	39	125,427	244,752	256,661	270,160
Fixed Platform	0-100	173	48,042	163,978	284,005	294,172
	101-200	140	56,922	224,423	281,210	322,129
	201-400	76	42,911	280,082	334,861	397,075

Footnote: (a) Includes 5 structures in the 200+ category

Table D.3

Average Revenue Threshold Levels At/Near the Year of Last Production – Lease Category I

Structure Type	Water Depth (ft)	n	$\overline{R}(t_{lp})$ ($)	$\overline{R}(t_{lp}-1)$ ($)	$\overline{R}(t_{lp}-2)$ ($)	$\overline{R}(t_{lp}-3)$ ($)
Caisson	0-100	140	598,846	1,414,225	1,953,555	1,752,930
	101-200	30	675,836	2,404,673	2,780,803	1,874,705
Well Protector	0-100	34	796,856	1,829,445	2,593,270	2,576,721
	101-200[a]	39	1,449,434	3,636,186	3,414,770	3,502,795
Fixed Platform	0-100	173	637,850	2,162,096	3,822,129	3,759,962
	101-200	140	739,295	3,069,187	3,971,520	4,630,609
	201-400	76	556,447	3,674,762	4,527,771	5,225,356

Footnote: (a) Includes 5 structures in the 200+ category

Table D.4

Average Production Threshold Levels At/Near the Year of Last Production – Lease Category II

Structure Type	Water Depth (ft)	n	$\overline{Q}(t_{lp})$ (BOE)	$\overline{Q}(t_{lp}-1)$ (BOE)	$\overline{Q}(t_{lp}-2)$ (BOE)	$\overline{Q}(t_{lp}-3)$ (BOE)
Caisson	0-100	376	30,959	96,127	142,145	179,518
	101-200	14	33,155	53,670	99,175	158,244
Well Protector	0-100	103	23,832	72,639	117,468	146,701
	101-200	20	67,355	212,480	329,858	539,504
Fixed Platform	0-100	107	25,197	82,540	104,389	154,998
	101-200	51	41,349	138,647	162,908	253,433
	201-400	15	89,712	478,356	527,775	611,107

Table D.5

Average Revenue Threshold Levels At/Near the Year of Last Production – Lease Category II

Structure Type	Water Depth (ft)	n	$\overline{R}(t_{lp})$ ($\$$)	$\overline{R}(t_{lp}-1)$ ($\$$)	$\overline{R}(t_{lp}-2)$ ($\$$)	$\overline{R}(t_{lp}-3)$ ($\$$)
Caisson	0-100	376	385,188	1,183,751	1,815,805	2,244,257
	101-200	14	457,331	785,508	1,225,973	1,932,231
Well Protector	0-100	103	261,355	884,105	1,554,026	1,812,339
	101-200	20	1,034,012	2,851,025	4,331,782	5,938,351
Fixed Platform	0-100	107	350,700	1,179,240	1,698,004	2,529,227
	101-200	51	675,103	2,069,231	2,509,057	3,605,908
	201-400	15	1,112,651	5,964,440	6,187,286	7,517,325

119

Table D.6

Average Production Threshold Levels At/Near the Year of Last Production – Lease Category III

Structure Type	Water Depth (ft)	n	$\overline{Q}(t_{lp})$ (BOE)	$\overline{Q}(t_{lp}-1)$ (BOE)	$\overline{Q}(t_{lp}-2)$ (BOE)	$\overline{Q}(t_{lp}-3)$ (BOE)
Caisson	0-100	75	36,682	182,195	243,703	236,718
	101-200	4	29,920	153,594	174,716	185,420
Well Protector	0-100	29	30,352	133,058	226,897	354,538
	101-200	5	3,076	37,216	63,760	81,609
Fixed Platform	0-100	36	41,678	153,705	279,942	583,044
	101-200	15	34,685	107,269	135,505	193,528
	201-400	3	30,176	622,324	435,370	130,086

Table D.7

Average Revenue Threshold Levels At/Near the Year of Last Production – Lease Category III

Structure Type	Water Depth (ft)	n	$\overline{R}(t_{lp})$ ($)	$\overline{R}(t_{lp}-1)$ ($)	$\overline{R}(t_{lp}-2)$ ($)	$\overline{R}(t_{lp}-3)$ ($)
Caisson	0-100	75	584,766	2,483,402	3,160,995	2,914,919
	101-200	4	348,021	1,825,733	2,205,311	2,635,662
Well Protector	0-100	29	420,353	1,865,306	3,497,969	5,498,238
	101-200	5	41,416	450,131	803,827	1,058,263
Fixed Platform	0-100	36	579,338	2,233,477	3,795,996	7,506,140
	101-200	15	569,607	1,730,695	2,411,147	3,523,914
	201-400	3	463,694	12,209,740	8,786,409	1,597,371

Table D.8

Average Production Threshold Levels At/Near the Year of Last Production – Lease Category IV

Structure Type	Water Depth (ft)	n	$\overline{Q}(t_{lp})$ (BOE)	$\overline{Q}(t_{lp}-1)$ (BOE)	$\overline{Q}(t_{lp}-2)$ (BOE)	$\overline{Q}(t_{lp}-3)$ (BOE)
Caisson	0-100	221	38,921	124,064	159,916	201,616
	101-200	2	68,053	223,518	430,632	478,033
Well Protector	0-100	30	22,832	87,589	105,091	133,597
	101-200	0				
Fixed Platform	0-100	30	37,454	134,922	201,851	279,217
	101-200	15	45,328	157,210	213,253	221,522
	201-400	1	10,183	67,671	96,914	94,853

Table D.9

Average Revenue Threshold Levels At/Near the Year of Last Production – Lease Category IV

Structure Type	Water Depth (ft)	n	$\overline{R}(t_{lp})$ ($)	$\overline{R}(t_{lp}-1)$ ($)	$\overline{R}(t_{lp}-2)$ ($)	$\overline{R}(t_{lp}-3)$ ($)
Caisson	0-100	221	527,742	1,558,099	1,933,574	2,237,764
	101-200	2	813,918	3,334,778	1,826,155	2,420,007
Well Protector	0-100	30	353,362	568,310	1,279,857	1,637,918
	101-200	0				
Fixed Platform	0-100	30	513,304	1,758,019	2,419,529	3,139,999
	101-200	15	614,269	2,082,317	3,003,751	3,042,329
	201-400	1	140,919	907,982	1,446,346	1,503,154

APPENDIX E
CHAPTER 5 FIGURES AND TABLES

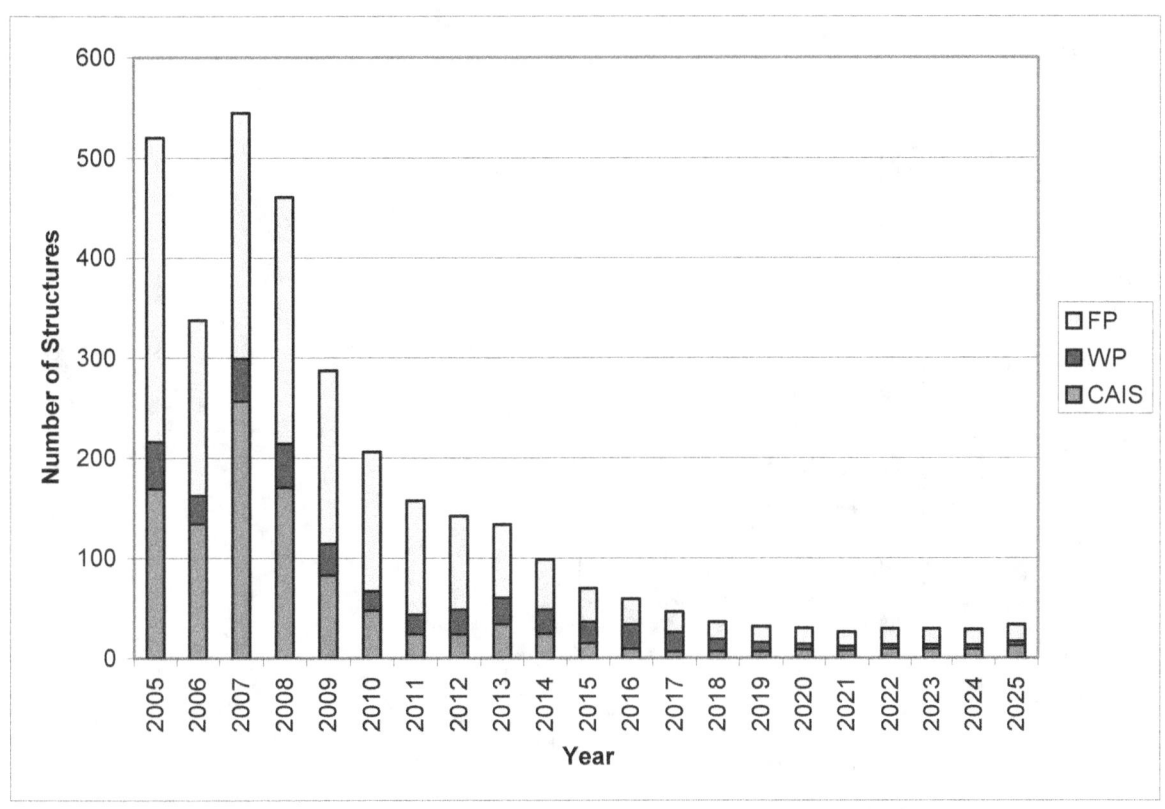

Figure E.1: Central GOM Production Threshold Structure Removal Forecast.

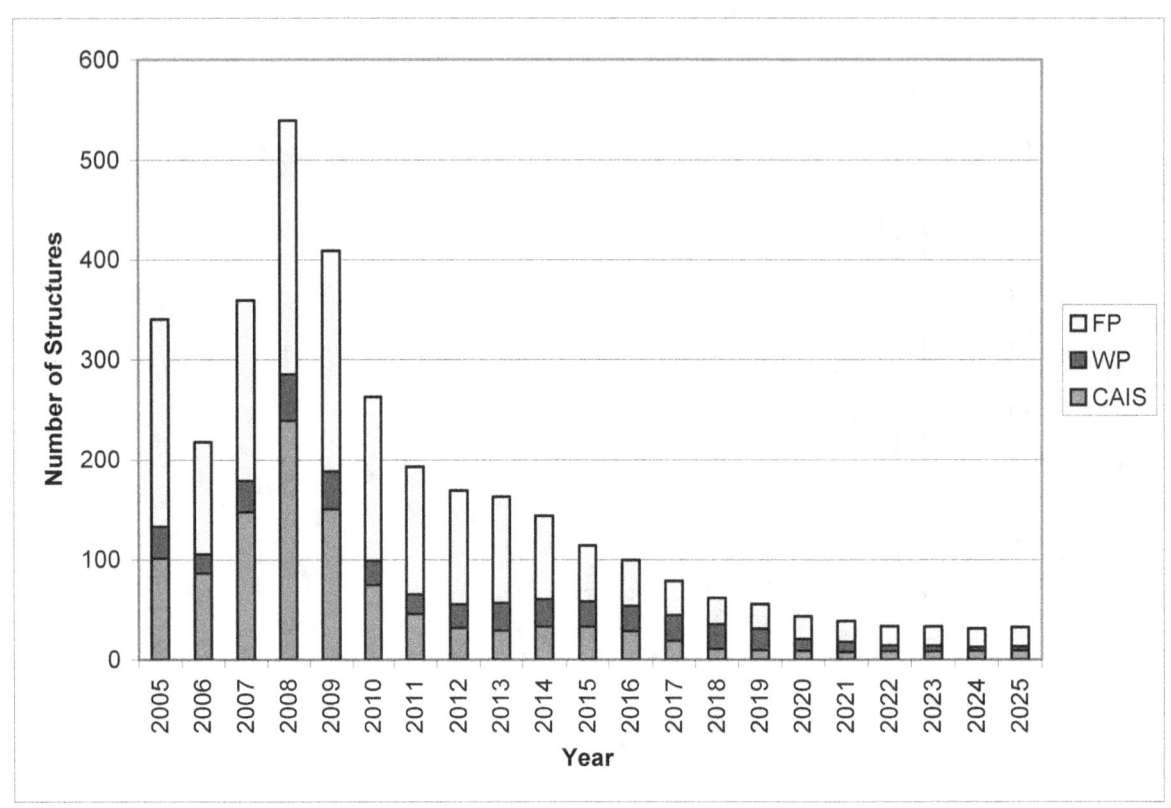

Figure E.2: Central GOM Revenue Threshold Structure Removal Forecast.

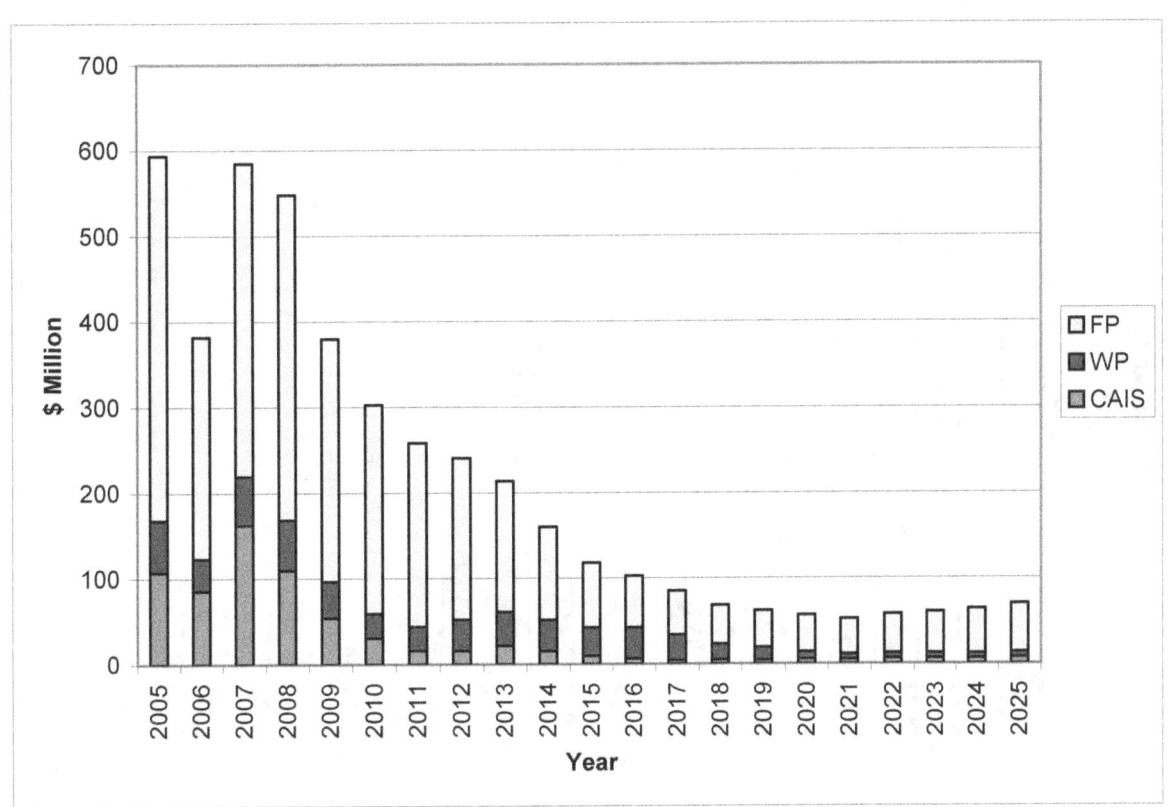

Figure E.3: Central GOM Production Threshold Removal Cost Forecast.

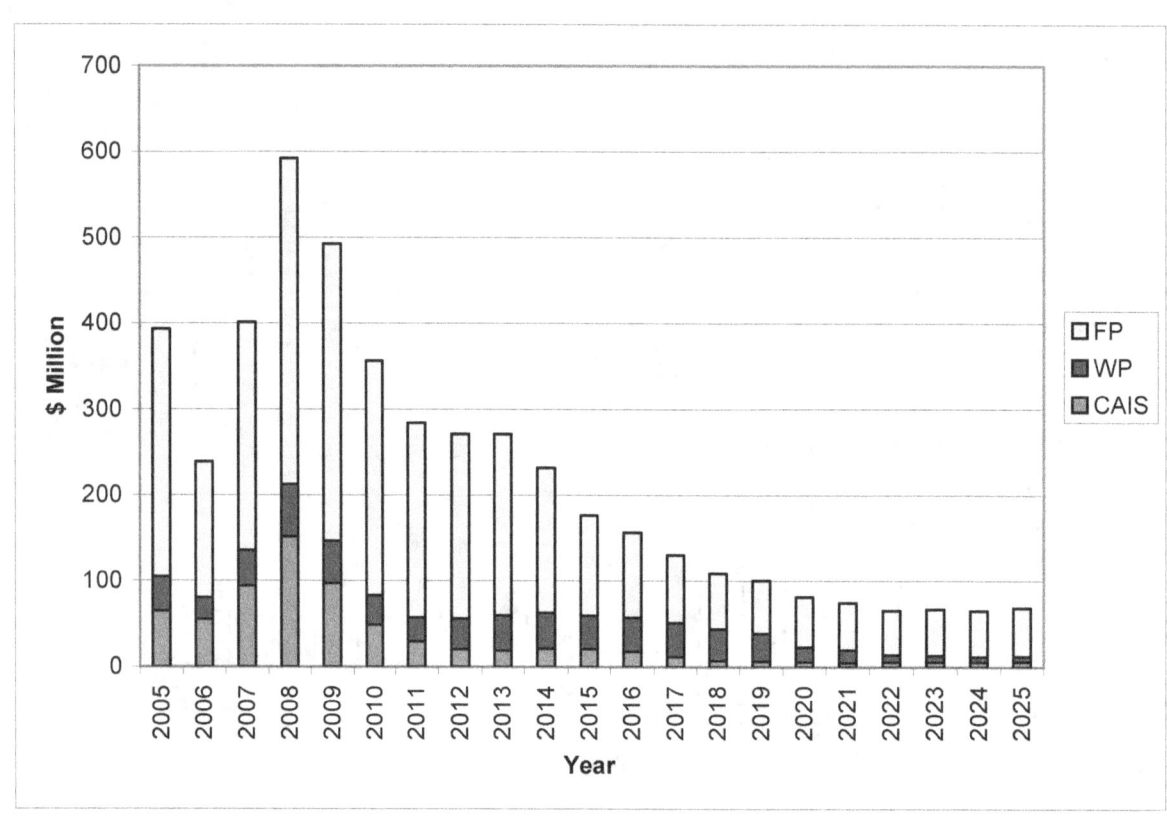

Figure E.4: Central GOM Revenue Threshold Removal Cost Forecast.

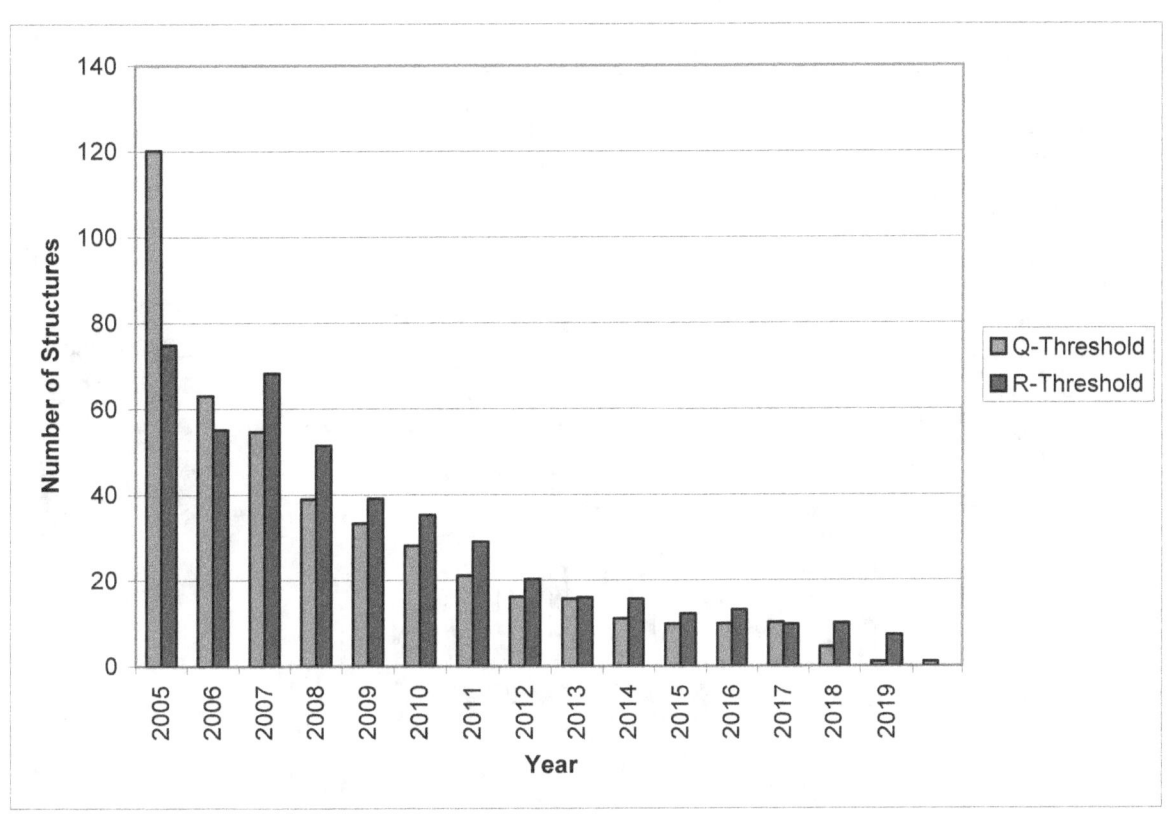

Figure E.5: Western GOM Structure Removal Forecast Comparison.

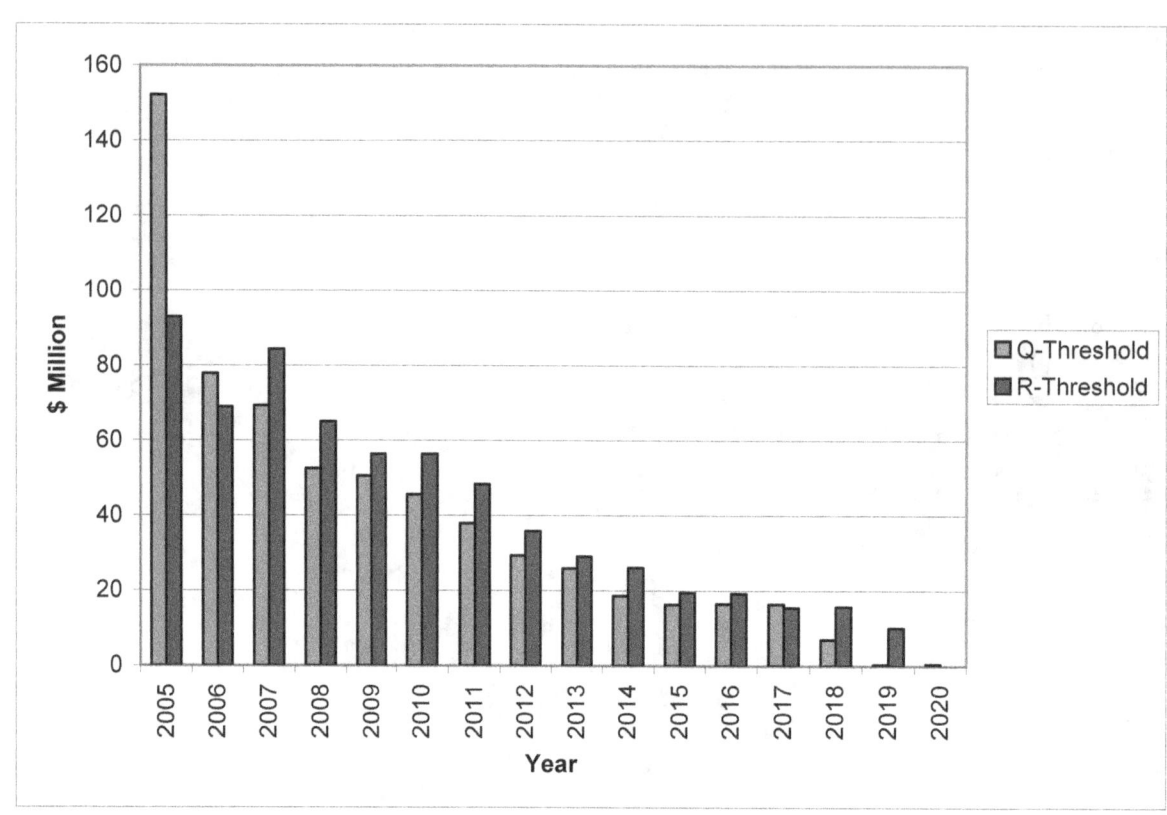

Figure E.6: Western GOM Removal Cost Forecast Comparison.

Table E.1

Number of Structures Removed in the Gulf of Mexico (1973-2002)

Year	Caisson	Well Protector	Fixed Platform	Total
1973	1	0	0	1
1974	4	1	0	5
1975	24	9	3	36
1976	20	8	2	30
1977	10	5	2	17
1978	18	3	5	26
1979	21	4	10	35
1980	19	8	9	36
1981	16	2	6	24
1982	8	2	5	15
1983	22	6	10	38
1984	25	14	14	53
1985	30	11	14	55
1986	16	8	10	34
1987	10	2	11	23
1988	55	8	36	99
1989	48	9	37	94
1990	60	11	37	108
1991	57	16	44	117
1992	48	13	45	106
1993	78	30	64	172
1994	43	16	66	125
1995	59	8	46	113
1996	49	15	55	119
1997	92	14	71	177
1998	36	11	29	76
1999	74	18	46	138
2000	52	20	69	141
2001	33	15	58	106
2002	24	17	54	95
TOTAL	1,052	304	858	2,214

Table E.2

Active, Idle, and Auxiliary Structures on Active Leases (2003)

k	Number of active leases with k active structures	Number of active structures	Number of idle structures	Number of auxiliary structures
1	944	944	291	129
2	245	490	141	79
3	84	252	96	66
4	35	140	84	43
≥ 5	48	348	286	123
Total	1,356	2,175	898	440

Table E.3

Active, Idle, and Auxiliary Structures in the Gulf of Mexico (2003)

Water Depth (ft)	WGOM			CGOM			GOM
	CAIS	WP	FP	CAIS	WP	FP	Auxiliary
0-20	1	0	0	200	10	35	79
21-100	79	25	119	767	268	710	318
101-200	3	17	82	48	63	490	73
210-400	1	4	85	1	12	320	31
400+	0	0	13	0	3	43	4
TOTAL	84	46	299	1,016	356	1,598	505

Table E.4

Idle and Auxiliary Structures on Inactive Leases in the Gulf of Mexico (2003)

Water Depth (ft)	WGOM			CGOM		
	CAIS	WP	FP	CAIS	WP	FP
0-100	22	7	29	103	20	88
101-200	3	4	6	13	10	44
210-400	1	1	9		1	30
TOTAL	26	12	44	116	31	162

Table E.5

Normalized Annual Production and Revenue Threshold Levels in the Gulf of Mexico

Threshold	Lease Categorization	Hydrocarbon Production	Water Depth (ft)	CAIS (MBOE)	WP (MBOE)	FP (MBOE)
Production	I	Oil	0-100	18	21	33
			101-200	30	30	66
			201+			34
		Gas	0-100	41	41	45
			101-200	51	47	52
			201+			51
	II	Oil	0-100	14	17	23
			101-200	20	20	44
			201+			25
		Gas	0-100	42	34	29
			101-200	41	31	31
			201+			36

Threshold	Lease Categorization	Hydrocarbon Production	Water Depth (ft)	CAIS ($1,000)	WP ($1000)	FP ($1,000)
Revenue	I	Oil	0-100	287	300	540
			101-200	512	608	623
			201+			595
		Gas	0-100	534	518	578
			101-200	631	640	697
			201+			695
	II	Oil	0-100	221	231	361
			101-200	488	614	806
			201+			545
		Gas	0-100	516	374	403
			101-200	546	744	478
			201+			1,172

The Department of the Interior Mission

As the Nation's principal conservation agency, the Department of the Interior has responsibility for most of our nationally owned public lands and natural resources. This includes fostering sound use of our land and water resources; protecting our fish, wildlife, and biological diversity; preserving the environmental and cultural values of our national parks and historical places; and providing for the enjoyment of life through outdoor recreation. The Department assesses our energy and mineral resources and works to ensure that their development is in the best interests of all our people by encouraging stewardship and citizen participation in their care. The Department also has a major responsibility for American Indian reservation communities and for people who live in island territories under U.S. administration.

The Minerals Management Service Mission

As a bureau of the Department of the Interior, the Minerals Management Service's (MMS) primary responsibilities are to manage the mineral resources located on the Nation's Outer Continental Shelf (OCS), collect revenue from the Federal OCS and onshore Federal and Indian lands, and distribute those revenues.

Moreover, in working to meet its responsibilities, the **Offshore Minerals Management Program** administers the OCS competitive leasing program and oversees the safe and environmentally sound exploration and production of our Nation's offshore natural gas, oil and other mineral resources. The MMS **Minerals Revenue Management** meets its responsibilities by ensuring the efficient, timely and accurate collection and disbursement of revenue from mineral leasing and production due to Indian tribes and allottees, States and the U.S. Treasury.

The MMS strives to fulfill its responsibilities through the general guiding principles of: (1) being responsive to the public's concerns and interests by maintaining a dialogue with all potentially affected parties and (2) carrying out its programs with an emphasis on working to enhance the quality of life for all Americans by lending MMS assistance and expertise to economic development and environmental protection.

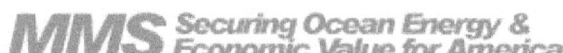